THE BEEKEEPER'S HANDBOOK

Gives practical advice for beginners, and is also a mine of detailed facts and theories for the more advanced. Includes sections on bee behaviour, anatomy, pollination, swarming and reproduction, queen rearing, honey harvest, diseases and pests, and microscopy and dissection.

D1264193

By the same author
BASIC BEEKEEPING

THE BEEKEEPER'S HANDBOOK

A Practical Manual of Bee Management

by

OWEN MEYER

THORSONS PUBLISHERS LIMITED
Wellingborough, Northamptonshire

First published in 1981

British Library Cataloguing in Publication Data

Meyer, Owen
    The beekeeper's handbook
    1. Bee culture
    I. Title
    638'.1   SF525

    ISBN 0-7225-0669-4

Typeset by Harper Phototypesetters, Northampton.
Printed in Great Britain by Whitstable Litho Limited,
Whitstable, Kent, and bound by
Weatherby Woolnough, Wellingborough, Northamptonshire.

Contents

		Page
Introduction		7

Chapter

1.	Making a Start	9
2.	Management	33
3.	Hives	39
4.	Handling Bees	59
5.	Behaviour	67
6.	Pollination	93
7.	The Anatomy of the Bee	97
8.	The Products of the Hive	117
9.	Swarming and Increase	143
10.	Queen Rearing	159
11.	The Honey Harvest	169
12.	Extracting the Honey	181

13. Diseases, Pests and Other Nuisances 195
14. Microscopy and Dissection 217
   Appendix 1: The Composition of Honey 231
   Appendix 2: Important Plants 233
   Appendix 3: British Legislation Affecting Beekeeper's 235
   Further Reading 239
   Glossary 243
   Index 249

Introduction

The success of my book *Basic Beekeeping* led to the suggestion that I should extend and amplify its content in a way which would cover most aspects of beekeeping in greater detail than was possible or desirable in the smaller book. What follows is my attempt to do this. It is based on personal experience and observation and any opinions expressed are my own. If I have given some stress to the Do-It-Yourself angle, even in such fields as anatomy and microscopy, it is because I am very much aware of the high cost of equipment and of the various bits and pieces which become needed as soon as the beekeeper decides to delve a little below the surface. Also, I am convinced that by making things by hand a much clearer understanding of the principles involved is obtained.

Beekeeping is not just an art, not just a science, not a hard-nosed business occupation and not just a collection of folksy traditions handed down from generation to generation. It is a curious amalgam of all four. The study of bees is not a static one either. Recent work on pheromones and trail markers, and electrical activity in the brain of the bee are examples of this, and although beekeeping is such an ancient occupation there is so much more yet to be discovered. I remember the late H. A. Dade saying to me that the study of the plant, fungal and animal occupants of a hive would be more than a lifetime's study for one man—even if the bees were ignored.

Over a fairly long period of beekeeping I have been privileged to meet and talk with many stimulating personalities and have been inspired by their example to do a little work myself—to my own very great pleasure. It is my hope that this book will encourage non-beekeepers to start with one or two colonies, and will suggest to more experienced beekeepers lines of practical research for them to follow. I know from personal experience that there is an enormous fund of expertise and know-how in the ranks of the ordinary beekeepers—some of them much better beekeepers than I. While this continues our Craft will prosper and the country benefit thereby.

As far as possible I have avoided the use of unnecessary technical jargon which experts in many fields of human endeavour tend to use as a smokescreen to make simple things appear complicated. Nevertheless, there are often ordinary words used in a beekeeping context which have a particular meaning for us—for example, the word 'colony'—and in other cases there is only one word available to denote a particular thing or process. Wherever possible I have explained these meanings in the text, but there is also a glossary for quick reference at the back of the book.

1

Making a Start

It is often said that the beekeeper's year starts in the autumn and I think this is also true of the very first year of the beginner. In September many Local Authorities and Beekeeping Associations start instructional courses for beginners and more advanced courses too for the beekeeper with a year or two of experience who wants to go deeper into the subject. Join one of these courses. Having thought about starting beekeeping you will certainly have read a selection of the many books published about beekeeping. You are likely to find that your lecturers will have notions of their own which do not always agree with what you have read but their basics will tally. This is all to the good. Beekeeping is an art not a precise science—although a good deal of science comes into it as you will see. As you progress you will find that you too will develop your own ideas to meet different situations. At the course you will meet other would-be beginners with whom you will be able to exchange ideas and you will be able to seek advice on specific problems from the lecturers.

In my firmly held opinion this is a good time to join your local Beekeeping Association. Names and addresses of Secretaries can usually be found in Public Libraries. It *has* been my ill luck to come across a few, a very few, badly run Associations with lethargic officers, but the vast majority are lively and well organized by enthusiastic Secretaries and you will get much gratuitous help and advice from officers and members alike. It is a curious fact that beekeepers by and large are friendly folk willing, even eager, to help a beginner in many ways.

Even if they do not run courses, Associations meet regularly during the winter months for discussion and talks by 'experts' (whatever these may be!) and in the active season meet in members apiaries where demonstrations of handling bees and examining colonies is the order of the day. All these activities involve a social aspect and what with this, insurance cover, the availability of advisory leaflets, magazines and the dissemination of up-to-date news about the latest developments, you will find that your small annual subscription is probably the most rewarding investment you have made for years.

Having spent the winter talking bees and reading about them, you will be anxious to start by having some of your own and 'setting up shop' by buying the equipment you will need.

Clothing

Before you get your first bees you should think about personal protection. A good veil is essential and should be worn on all occasions when handling bees or attending demonstrations with live bees. Gloves are also advisable. I fear that most of us are guilty of disregarding this cardinal rule sometimes, especially when we get to know the temper of our own bees. It is wrong to do so, nevertheless.

Head Gear

There is a choice of styles available for sale or you can make your own. A veil must be bee-tight and fit closely over the shoulders, back and chest and have some device to keep the front away from your face.

To make your own is simple. Buy a length of black nylon netting large enough to be made into a tube a little less than the width of your shoulders in diameter. Hem one end, sew into a tube and thread a length of elastic through the hem so that it will grip a hat tightly. An old straw hat is ideal. Sew the centre of a length of tape to the middle of the bottom of the back of the netting. In use the tape is passed under the arms, round the chest over the veil, back again and tied in a bow at the rear. Sew a ring of wire inside the veil at about mouth level. This will keep the netting away from your neck and face.

Gloves

Gloves give confidence and are recommended for the beginner. As confidence and expertise grow you will probably discard them, but I think there is a case for wearing gloves most of the time. There are many kinds of household and gardening gloves that can be adapted for our purpose. Some are a little clumsy, but some have the advantage of wide wrist parts to which can be attached cotton gauntlets elasticated at the top to reach up the arms. My own preference is for the purpose-made gloves made of thin kid leather with attached cotton gauntlets into which sleeves can be tucked. I like mine to be a fairly good, tight fit to make them less clumsy. I also like to have my forefingers and thumbs free so that I can pick up queens or other individual bees delicately. I have therefore cut off the ends of the glove fingers and thumbs so that the top joints of my forefingers and thumbs on both hands are bare. As my gloves fit fairly tightly no gaps are left up which bees can crawl. They are bound to do this if they can find a dark tunnel.

I find that I wear gloves fairly consistently now, except for brief examinations. There are two main reasons for doing so. I see no reason to be stung needlessly. It always hurts, even when one has become inured and has no bad reactions. Also, every bee which has used its sting is a dead bee. If I am wearing gloves and the bees turn a little 'niggly' I can carry on quietly with what I am doing without having to stop, look around for my gloves and put them on while the hive is open and the bees are getting more fractious. The second, and more important reason is that for some years now I have had no bees in my home garden apart from the odd nucleus. All are in out-apiaries a car journey away. Propolis and honey gets quickly transferred from hands to steering wheel, brake handle, door catches, switches and all manner of possible (and some impossible) surfaces. If I wear gloves I can peel them off when I have finished and drive home

Figure 1. Prepared for the worst: veil, trousers tucked into Wellington boots, leather gloves with cotton gauntlets reaching to elbows.

with clean hands. Most out-apiaries are sited where there are no washing facilities. Also, in the course of time the gloves become impregnated with the smell of propolis and honey and the bees do not seem to dislike this.

Further Protective Clothing

A pair of light-coloured trousers and an old linen jacket are quite good for doing your beekeeping in, but it is worthwhile giving a little thought to something more practical.

Bees dislike hairy woollen clothes and socks. Neither do they like certain colours, principally blues and browns. Light colours are better and white is best of all. They will often take a dislike to nylon. I believe this to be because of the static electricity generated by movement. Some people are more prone to generate an electrical charge than others. I suppose silk would act similarly, but I cannot imagine many beekeepers working their bees while clad in silk blouses or shirts! Propolis and honey will get on your clothes as well as on your hands and I think the ideal is some kind of white cotton overall garment. I use a cotton 'romper suit' which covers me from neck to ankles and has plenty of pockets for matches, spare smoker fuel and other odds and ends. It is really a decorator's overall and even has a narrow pocket down the right thigh on the outside intended for a folding rule. This just fits my hive tool. My only grumble is that working in overalls and a veil can be very hot. Still I do not *have* to wear woollies underneath and find that a minimum of light-weight clothing is perfectly satisfactory.

Essential Equipment

Hive Tools

A hive tool is merely a lever for separating frames and boxes which the bees have stuck together with propolis. They can be bought or made. Those offered by the manufacturers differ in design but all are excellent. I do not think there is any 'best' design. What is best for you is what you have become used to. You can even buy them made of stainless steel if you are so inclined (or have a convenient birthday). Whatever type you decide on it is a good idea to paint the whole of the middle part bright yellow or red. Hive tools have a way of laying themselves

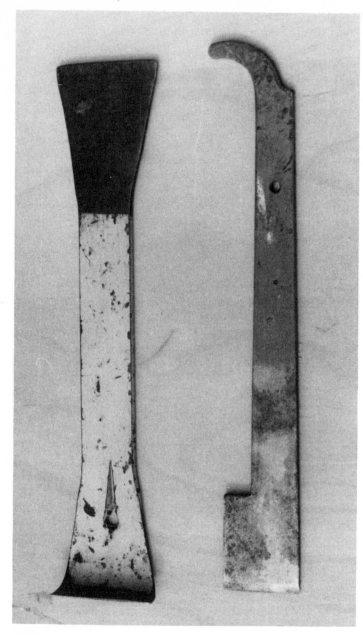

Figure 2.   Two types of hive tool, both of steel.

down in long grass. A veil impedes vision to some extent and a brightly-coloured hive tool is more easily seen.

Smokers

You will need your own smoker. Get a fairly large one. The old straight-nosed type has too small a fire box and the fuel will quickly burn through. The bent-nosed type is much better. The fire box will hold enough fuel for an hour-and-a-half or two-hour's burning and this will be long enough for the beginner. Some of the larger kinds will burn for much longer than this, of course. The professionals use a jumbo size smoker which will burn all day with one filling.

Smokers are made of tin plate or copper. The latter are considerably more expensive but will not corrode or burn through as the tin plate ones will eventually. Here again there is an opportunity for the handyman, especially if you are lucky enough to come across some sheet copper. I like to screw a picture hook to the back of the bellows so that I can hang the smoker on the edge of an open hive and leave both hands free.

What Sort of Hive?

The decision as to which type of hive you are to decide on is important. Whichever it is, adhere to the same type as your beekeeping expands. There is nothing more infuriating than having a mixture of different types of hive in one apiary, none of the parts being interchangeable. Some manipulations are not possible unless you have standardization of hives.

Consider what you have read about the various designs available to you and compare their good and bad points. Talk to local beekeepers in your association and find out what their preference is. This may well be influenced by the characteristics of the local strain of bee for, as we shall see, there is considerable variation in these.

I prefer the single walled type of hive. One additional advantage is that they lend themselves readily to home construction. Evening Institutes frequently have woodworking classes where expert advice and guidance are to be had. Only simple carpentry skills are needed, but it is imperative to work to standard dimensions. The odd plus or minus a quarter of an inch just will

Figure 3. Smokers. Left, a bent-nosed smoker, home-made from copper sheet—an old friend and practically indestructible. Right, a straight type with a smaller fuel box.

not do—it will lead to disaster. Working drawings of some of the more popular hives can be bought from the British Beekeepers' Association. These are clear to follow and metric and imperial dimensions are given together with a cutting list.

Secondhand hives in good condition can be bought and here your Beekeeping Association can be helpful. Frequently there is a member who, perhaps through advancing years, wishes to cut down his beekeeping and has surplus equipment for sale. Let the Secretary know your wants. He will be able to bring the prospective seller and buyer together.

All secondhand equipment should be sterilized—even if coming from a known and reliable source. Scrubbing with strong hot solution of washing soda is advised as one way, but I think the most positive and least messy way is to 'flame' everything with a blow torch, paying special attention to all nooks and crannies. Be careful of zinc excluders however. Zinc has a low melting point!

Personally, I have never used secondhand frames with drawn comb. Perhaps it is over-cautious, and I know they can be sterilized with the fumes of glacial acetic acid, but I feel that the risk of disease is too great. Any combs I have acquired have always had the wax rendered down in a solar wax extractor and the frames burnt.

Finally, of course, new hives can be bought from one of the Appliance Manufacturers. They are usually well-made and are, or should be, to standard specifications. Well-matured timber is hard to come by and shrinkage and warping may possibly be experienced. However, this possibility applies to home-made equipment too unless you can find a source of old planks in good condition. Floorboards are a case in point. They will be too narrow for you so some joining and preparation on both sides will be needed. Good old matured deal will make a rather heavy hive.

As soon as you start in earnest it is a good thing to acquire a second hive as an insurance against total loss and the provision of the possibility of increase. These chances always seem to come when you are least prepared for them. The aim should be two hives of similar design and, if possible, a nucleus box. The latter need be no more than a half-sized hive.

Getting the Bees

Since the middle of the nineteenth century bees, particularly queens, have been imported into this country from abroad, from Italy, France, Holland and more recently with the advent of air freight, queens of Caucasian and Carniolan ancestry have come in from Australia and the United States.

In the early years of this century our indigenous poulation of North European bees was swept by epidemic disease with devastating results making re-population necessary. The result of these two factors is that our present bee population is composed of a glorious mixture of North European, Italian, Carniolan, Caucasian etc. strains. I very much doubt if any British black bees have survived in the pure state although some beekeepers claim otherwise and say that pockets do exist in isolated areas. All these strains belong to one species, *Apis mellifera* and all have their good and bad points.

The old black European bee was economical with stores, wintered well, was not prolific and maintained a small brood nest so that a colony could be accommodated in a single brood chamber. It made beautiful white cappings on honeycomb. It was inclined to be sharp tempered on occasions and ran about on the comb during examinations.

Italian strains are prolific and need more space than is provided by a single brood chamber and want ample stores on which to winter. They are gentle and quiet on the comb and the queens are large and light-coloured so they are easy to find.

Carniolan bees are gentle and quiet on the comb. They winter with small colonies and consequently need smaller winter stores. They use little propolis and have a reputation for being comparatively free from brood diseases. They have a strong disposition to swarm.

Caucasians too are gentle and quiet on the comb. They build up strong colonies but do not usually reach a peak until mid-summer. They do not swarm readily but use large quantities of propolis. A favourite habit is to build a wall of propolis just inside the entrance on the floor leaving only a small exit hole. They are susceptible to Nosema disease. The proboscis is longer than that of the other races.

The above is only a very brief summary of the more obvious

characteristics of the different races with no reference to the differences in outward appearance. What we are concerned with here is the potential of the bees available in our mixture which may be of use to us as practical beekeepers.

Breeding Bees

Attempts have been made to breed a 'better bee' and probably the most successful and certainly the best-known has been the Buckfast strain. Brother Adam of Buckfast Abbey is a beekeeper and a bee-breeder of world renown and has rightly earned the respect of beekeepers from all quarters. He has travelled extensively throughout the civilized world (and a good deal of the uncivilized too) seeking out native strains of bees, assessing their suitability as breeding material and trying with considerable success to accentuate the desirable qualities and breed out the bad. In order to do this he has had to carry out some degree of inbreeding. This would have inevitably led to a loss of vitality if carried to extremes but Brother Adam is too good a geneticist to have allowed this to happen. In half a century and more of careful selective breeding he has produced a cross of distinct races which has hybrid vigour and many good characteristics.

One essential of line breeding is the establishment of mating apiaries in isolated situations such as the Buckfast mating stations in lonely parts of Dartmoor. Only by doing this can the breeder be quite sure that the only drones flying will be those of his own choosing. Instrumental insemination too has been used at Buckfast. This has limited application but is of inestimable value in certain circumstances. There can be no true breeding unless there is complete control of both parents. This most of us do not have.

The ordinary beekeeper is not able to keep his bees in isolated places and has little or no control over the drones in the vicinity of his apiaries or mating stations and the vast majority of our bees are, to put it impolitely, mongrels with as mixed a bag of ancestry as you could hope to meet. This is one reason why buyers of Buckfast queens, or other line bred queens for that matter, are often delighted with their purchases but very disappointed with the first cross from them.

Ancestral characteristics, recessive in the drones but dominant in their offspring give rise to undesirable traits. The most noticeable of these are bad temper and an increased tendency to swarm.

What good beekeepers do—and what all beekeepers might do—is to forget about 'breeding' in the proper sense of the word, but to raise queens from their best stocks and to cull the worst. This should be a continuing process. Daughter queens from a known good stock do not always run true to type. There is an element of luck in random matings and hybridization, or *heterosis* as the geneticists call it, does not always produce the results desired. However, by continued selection in this way the general quality of the bees is bound to rise. I do not think the importation of bees from abroad is either desirable or necessary, quite apart from questions of legal bars to imports. We have imported disease and trouble for too long in the past while we have plenty of good strains of our own.

Evolution works by natural selection—but slowly. Perhaps we could give Nature a little help and accelerate the time scale by doing a modicum of selection ourselves. I am convinced that if this were done generally, in two human generations we would have a much more pleasant and productive bee population.

Sources of Bees

The preceding considerations lead me to suggest that when you are ready to get your bees (or increase your stocks) one of the best ways is to go to a local beekeeper who has a good reputation (as a beekeeper) and is known to have good bees and see if he will sell you a nucleus with a young queen. Many beekeepers make a sideline of producing nuclei. Critics will say that in the United Kingdom it is not possible to get nuclei 'early enough'. Early enough for what? Beginners hope to become and remain beekeepers for many years and although May is a pleasant time to start, the only disadvantage of starting with a nucleus in the summer is that you may not get much of a honey crop in your first year. Even this need not be the case if you are lucky enough to have a good hot summer.

This is another occasion when talking with members of an Association will point you in the right direction. Satisfactory nuclei can also be bought and the bees are almost invariably gentle to handle—the suppliers are likely to have carried out their own selection plan if only with an eye to future business. These nuclei come in non-returnable boxes which can after-wards be used as swarm boxes or even as temporary nucleus boxes if provided with a roof. Perhaps this is the place to say that in my experience the commercial firms are always helpful and pleasant to deal with. Most have been in the business for generations and seem to be staffed by people who know their business and will willingly help with queries and advice.

The cheapest way of all to start is to find a swarm hanging on a convenient bough, take it and hive it. Associations will often maintain a list of members wanting swarms so that news can be passed on. I do not recommend this method. One seldom knows where the swarm has come from, the history of the parent colony or whether disease is present. This is quite impossible to diagnose in a swarm. Swarms are best left until you have a little more experience and if possible have an isolation ward site where you can hive swarms and keep a close watch on them for a while.

The Bees Arrive

Let us assume that you have bought a nucleus from either a local beekeeper or from a bee-breeder through the trade channels, although initial handling will be much the same if you have bought a full-sized colony. The bees will arrive in either a nucleus box or a travelling box with a gauze screen in the lid for ventilation. If the bees are coming by rail, contact your nearest railway station, let them know that a box of live bees will be consigned to you in the near future and arrange for you to be telephoned as soon as they arrive. Railway staff are pretty good about live animals and in any case will be quite glad to get rid of a box of live bees. You will then be able to collect them with a minimum of delay.

By this time you should have set up your hive on the site where it is to remain. Remove the roof, brood boxes etc., but leave the floor where it is. Set down the nucleus box/travelling box on the hive floor with its entrance hole facing the way your hive entrance will. Open the entrance to allow the bees to fly and leave them to settle down. If the bees have arrived early in the morning—say before noon, the next operation can be carried out in the evening at about half past five or six o'clock. If they have come late in the day, I should leave them overnight and put a waterproof cover over the gauze screen. In the evening (or next day when the sun is well up, as the case may be) go to the bees dressed in your protective clothing. Take with you your smoker charged with fuel and well alight, your hive tool, two frames of foundation, a feeder full of sugar syrup and two dummy frames. These are simply normal frame top bars of the length suitable for your chosen hive but have a plain wooden board instead of the usual side and bottom bars filled with foundation.

Blow a *gentle* puff of smoke across the nucleus entrance and across the gauze screen. You must now wait a little for the bees to react to the smoke. They will not be alerted in an aggressive way but will start to gorge themselves with honey. While you are waiting, quietly lift the nucleus box off the floor board and put it down next to the hive. Place an empty brood box on the floor with an entrance block in position with the smaller entrance in the operating position. You will be helping your bees to defend the home by giving them a small exit and entrance to defend. Remove the lid of the nucleus/travelling box and any inner cover there may be and blow another gentle puff of smoke across the tops of the exposed frames.

Transfer each frame from the box to the brood chamber of your hive being very careful to keep them in the same order. Take your time over this and try to keep all your movements smooth and deliberate with no jerks. You may need to use the hook end of your hive tool in order to free the frames. It is hard not to look for the queen while you are transferring the frames. Keep an eye open for her by all means, but do not make a prolonged search. It would be much more profitable to look for and estimate the amount of stores of honey and pollen and also for the appearance of sealed and unsealed brood. When you have transferred all the frames you will probably find quite a few bees left in the box. Invert this over the hive and shake the bees into it. A good downwards shake coming to an abrupt stop will shake them down. Take care not to bang the box down on to the hive. See that the frames are pushed tightly together and add a frame of foundation at either end followed by the dummy boards. If you have only one dummy you can push the frames hard against one side wall of the brood chamber and finish with the single dummy board on the outside of the last frame of foundation. Place the inner cover on the brood box with the full feeder over an open feed hole. A shallow super is now needed to allow space for the feeder. Then replace the roof.

Leave well alone for a week other than to see that the feeder is topped up and to watch what is going on at the entrance. Try to resist the temptation to open up the hive 'just for a minute to see how they are getting on'. In fact they will get on much better without this sort of disturbance.

After seven days the bees will have settled down and you can make an examination of the combs. Again it is not absolutely essential to see the queen. The sight of eggs and very young larvae should be sufficient evidence that you have a laying queen at work. I think that often colonies of bees are pulled about unmercifully to find the queen even when the beekeeper has already seen plenty of evidence of her presence. Space, food stores and healthy appearance are much more important factors worthy of your attention. Of course there are occasions and operations when it is vital to find and secure the queen.

Finding the Queen

As we have noted, there are occasions when finding the queen is a necessary preliminary to some special operation and it is useful to be accustomed to her appearance and where she is likely to be found. There is no royal road to finding her (no pun intended). It is a knack which most, but by no means all, beekeepers acquire without quite knowing how. If you are specifically looking for the queen, use little smoke and disturb the colony as little as possible. She is easier to find if the bees are quietly going about their business and not rushing about the combs madly in a disorganized way. Some strains of bees are better than others in this respect. They remain quiet on the comb and are therefore much more pleasant to handle.

Remember that the queen has a longer abdomen than the workers, carries her wings folded neatly back and not protruding at the sides and has longer legs so that she stands up a little over the workers if the comb is viewed side-ways. She walks about in a more deliberate way and is surrounded by a ring of workers facing her. She is to be found in an area of the brood nest where there are eggs or patches of empty cells polished for the receipt of eggs. She will prefer darkness to light and will not be found on combs of sealed stores, unless you have driven her there.

With these points in mind, when you take off the inner cover, look down at the seams of bees between the frames. The queen is likely to be where there is the heaviest concentration of bees. If you smoke heavily you will disturb the pattern of their movements. For those interested in one-upmanship it is sometimes possible to get a totally undeserved reputation by pointing to a couple of frames in what you think is the brood nest and saying to the onlookers 'I think the queen is about *there*'. If, in the event, your prognosis is proved right your audience will think what a good beekeeper you are. If the queen is found on quite a different comb the drill is to say that she has been disturbed a little and has moved over. Of such is the mendacity of beekeepers!

But back to practicalities. One at a time, take out the first two frames from the side nearest to you. Examine each carefully and make quite sure that the queen is on neither. This is most unlikely as these frames will be full of stores. Hang the two frames in an upturned roof or a spare brood chamber. Now carefully go through all the remaining frames looking for her and leave the examined frames hanging in pairs in the brood chamber with a space between each pair. This is why you have taken out two frames to start with.

The idea of hanging the frames in pairs is that if you cannot find her on the first time through she may well be between a pair of frames in the comparative darkness. If you have your bees on a double brood chamber system, by inserting a queen excluder between the two boxes before you start looking you will divide the hive into two separate boxes to look through and at the same time prevent the queen running up (or down) into frames you have already examined.

If you cannot find her on your second time through then, apart from a third try, there are three courses open. You can close up the hive and try again another day, probably the wisest choice; you can move the hive some yards away, leave an empty brood chamber on the old site and then go through again. The flying bees will go back to the old site and leave you with less bees to deal with. Finally, you can filter the bees through a queen excluder on an empty brood chamber with another box on top of this to receive the bees from each comb you shake into it. Smoke the bees down and the drones and the queen will be left on the top of the excluder. This really does stir the bees up into a disorganized state and is not recommended, except in cases of dire urgency.

Do not forget to look at the floor and walls. Some queens tend to get there if chased.

Handling and Marking the Queen

When a queen is in full lay the ovaries are greatly expanded and

Figure 4. How to hold a queen. First, pick her up by the wings with the forefinger and thumb of one hand. Second, with the forefinger and thumb of the other hand, hold her firmly but gently by the legs. In this position she will lie comfortably in the little valley between your thumb and forefinger, and can then be marked or clipped.

occupy practically the whole of the abdominal cavity. Since she is fed entirely on royal jelly her digestive system does not have to deal with solid matter, the ventriculus does not have to expand after eating. Indeed, on dissection, the ovaries pop out immediately the dorsal plates are removed and this seems to indicate that the ovaries are actually under pressure. This is the delicate part of the queen and should be handled with great care or preferably not at all. When picking a queen up, do so by gently holding the wings between the forefinger and thumb of the right hand (or the left if you are left handed), then hold her legs between the forefinger and thumb of the other hand. You can then mark or clip her without there being any pressure on the abdomen. Some people prefer to hold her head and thorax between the tip of the thumb and the first two fingers, allowing the abdomen to rest on the thumb. A marked queen is always much easier to find and this is well worth doing. There is an international colour code for marking queens to indicate the year of mating, and this is shown below.

Table 1.   International Colour Code to show Year of Mating of Queen Bee.

Year ending in	Colour
1 or 6	White
2 or 7	Yellow
3 or 8	Red
4 or 9	Green
5 or 0	Blue

However, a spot of white on the thorax is readily visible and is quite adequate for most people who will know the ages of their queens anyway. Apart from making her easy to find marking has one other advantage—as an indication of supersedure. If you have marked your queen you will know that the bees have quietly replaced her by supersedure if you find an unmarked queen at a later inspection.

In spite of the need to handle queens carefully, she is not a feeble insect in any way and is quite as strong as the workers. If you keep clear of her abdomen there is no need to be excessively cautious otherwise. She is quite robust and will never sting you. She seems quite devoid of any aggressive instinct and her sting will only be used on other sister queens at the time of hatching.

To mark her, use a quick drying paint. Put a little pool of paint on a convenient horizontal surface and dip the end of a match into it (not the striking end), dab the painted end of the matchstick on to the queen's thorax and give it a little twist between finger and thumb. Do not grind it down—just a gentle twist will do the trick and will leave enough paint behind to act as a marker for you when you need to find her. Let her run back on to the comb on which you found her or on to a comb of brood.

Clipping the Queen

This is a device intended to prevent swarms escaping. It is claimed that a clipped queen either cannot fly at all or, at best, can only fly a short distance, and then in a lopsided fashion. About half of each forewing can be cut off or (more usually) about one third of one wing. A small pair of sharp scissors is the tool for this job and the operation is usually done early in the queen's life—say at the nucleus stage when she has commenced to lay. Care must be exercised to see that she does not slip a leg between the blades of the scissors just before they are closed. This she can easily do especially if she is struggling a bit to get out of those clumsy hands holding her. There are no nerves in the wings and cutting seems to cause no pain or inconvenience except in flying. What are sometimes called 'nervures' in the wings are not nerves at all but merely stiffening structures.

In some strains of bees supersedure will often follow wing-clipping. Could dislike of mutilation be the cause?

The Season Progresses

Your nucleus will grow and expand sideways quickly enough on spring flowers and you will get much satisfaction from seeing those nice khaki slabs of sealed brood. These are your own bees. You may have bought a nucleus of bees raised by someone else but these combs of bees have been bred by you and really are *your* bees. As the spring blossom appears in quantity you will be able to discontinue feeding syrup too, but beware of those cold spells and keep a watch on the food situation. Before long you will have a full-sized colony expanding to cover eleven frames. It is always a good thing to give space in advance of need and before the bees occupy the whole of the brood chamber you should do this by adding a super of shallow frames over an excluder. By the end of May you should have one super on.

At this time of the year when breeding is in full swing, bees consume a considerable volume of water. Some they will get from raindrops or dew on plant leaves but this is not enough and bees can always be found drinking from water sources—sometimes not too savoury ones either. Their predilection for pools of rainwater on cow pats can be obviated to some extent by supplying them with a source of clean water. A plastic bowl or an old stone sink half-filled with pebbles so that the bees cannot drown will do well enough. Keep this close to your hive(s) and keep it topped up to the pebbles with clean water.

Colony Development

The growth of a colony and its development into a balanced unit is controlled by the time of the year, weather conditions and the availability of nectar and pollen. The bees themselves too have a biological rhythm—a kind of biological 'clock' which seems to operate independently of the external environment. The first three factors are obviously closely interrelated and variations in the norm of weather conditions will affect the development of flowers. Not only current conditions are responsible. The previous year's rainfall and temperatures will have affected the subsoil so that there is always a 'carry over' effect. Apples and clover are only two examples of this—and for different reasons. In apples, growing conditions in the year before fruiting have a powerful effect on the future crop. In the case of wild white clover it is the temperature and moisture content of the subsoil, the result of the previous year's weather, which have a strong influence on nectar yield.

In spite of all these imponderables, it is possible to formulate guide lines of average colony development. The chart attempts to show colony development in an average year in conditions which might be expected in the centre of the United Kingdom.

COLONY DEVELOPMENT

rate of egg laying

foragers

COLONY INTAKE

pollen

nectar

	FEB	MAR	APL	MAY	JUN	JUL	AUG	SEP	OCT	
		Crocus Hazel Alder Willow	Wallflower Dandelion Cherry Blackthorn	May Sycamore Apple	minor sources	Clover Blackberry Mustard Lime	Willowherb Mallow Sweet Chestnut	Golden Rod Mich. Daisy Balsam	Mich. Daisy Ivy Thistle	
	FEB	MAR	APL	MAY	JUN	JUL	AUG	SEP	OCT	

Chart showing Colony Intake and Colony Development

There will, of course, be variations in other districts, some quite considerable. The sources of nectar and pollen too are indications only. There are many other sources of forage of importance. There is one big omission. The chart shows nothing of the effect of the heather honey crop.

If we accept the implications of the chart then it is clear that only the eggs laid from mid-May to mid-June will produce the foragers to bring in the main July nectar flow, and this can be of short duration. The second peak of egg laying stimulated by the main flow will produce a generation of foragers available for the heather crop later in the year for those who take their bees to the heather. The 'mini' flow in September and October is a fortuitous gift of Nature to stimulate egg-laying so that there will be plenty of young bees, not worn out by foraging, available for wintering.

A steady intake of nectar and pollen over the period March to May is necessary for a gradual egg-laying build up, but un-

Figure 5.   Spring examination—a second cycle of brood.

fortunately in some years bad weather precludes this and resort has to be made to feeding sugar syrup. Most beekeepers have had the experience of bees being on the point of starvation in June in exceptionally bad years. The moral is to keep a close eye on the food situation at all times but especially in the early part of the year because this is when the food-needs of the colony are increasing rapidly. Brood cannot be raised successfully without an adequate supply of fresh pollen.

If it were possible it would be best to leave the bees unmolested in the periods May to mid-June and for the whole of July so that the queen can get on with her business of raising brood in the first period and the foragers can concentrate on the nectar flow in the second. Unfortunately, this is the period when we can normally expect to find preparations for swarming. To be sure of not losing valuable swarms, inspections for swarm cells should be made at weekly intervals as suggested in the section on 'Routine Examinations for Swarming' (p. 155). However, these examinations need not be prolonged nor need they be too disturbing to the colony.

If preparations for swarming are found, as we will see in the 'Swarming' section, two choices are open—either to let things take their course and to let a swarm issue, hoping that you will be about at the time to secure it. This course is not to your advantage and in some circumstances could be regarded as anti-social. Or the second choice is to put into operation whatever swarm control method you fancy even if this is no more than repeatedly cutting out queen cells every seven days. I do not like this method very much. Swarming, the urge to procreate, is so strong and fundamental that I fancy the bees eventually become demoralized and lethargic. There is the possibility that you may miss one queen cell—and one is enough for a swarm. In a demoralized state the colony will have lost much of its energy and will be little use as a nectar gathering force for some time.

On the whole the best method for the beginner is some kind of artificial swarm. There are many variations on this theme involving the switching of flying bees from colony to colony, but in the early days it is best to keep it simple. Your artificial swarm need be no more than the removal of the queen on her comb with two frames of food to a nucleus box or even a spare hive a few yards away and the cutting out of all queen cells in the parent stock except one good-shaped unsealed cell. See that you have not taken a queen cell with the queen in the nucleus.

A word of advice and a little supervision from an experienced beekeeper will help you over your first hurdle, but the operation is basically simple and well within the competence of the tyro. Leave both lots alone and do not feed the nucleus for at least a week in order not to encourage robbing. You will find that you have allowed the bees to work out their swarming impulse, not frustrated it. You will have lost no bees but will have gained a useful nucleus which you can build up into a full colony or, if you do not want increase, you can unite it back to the parent colony later in the season after killing the old queen. Since you have kept all the foragers on the old site, their nectar gathering capacity will be unimpaired.

Apart from these inspections for signs of swarming leave the colony as far as possible undisturbed until it has gathered the main nectar flow. This will mean seeing that it has plenty of empty super room at all times. It is surprising how quickly a good colony will fill a super at the onset of a nectar flow. If you decide that you want to increase your stocks, it is well to remember that you can have more bees *or* a honey crop, but not usually both. The secret of a good honey crop, given good weather, is to have a powerful force of foragers in good heart at the time the main nectar flow is normally expected in your area.

This is not to say that taking out a nucleus from a strong colony cannot be a useful practical operation—it can. It can be a safety valve in a pre-swarming situation, but it must be done judiciously, as early as possible in the year and with a strong colony.

2

Management

In beekeeping, by management we mean the adoption of a system of control which will lead to the maximum possible production, whether it be of run honey, comb honey and sections, heather honey or of colonies of pollinators, strong and ready to fulfil pollination contracts. Not all these objects are mutually compatible, but the problem can be broken down into five main areas for consideration.

Forage

Each apiary should have an abundance of nectar-secreting and pollen-bearing plants in the near vicinity. Although bees will forage at a distance from the hive it is little use relying on nectar sources at or near the limit of their pay load flight. The beekeeper should be aware of the period and duration of the expected main flow and of minor flows too, which are so useful in the building up period. The object of his management should be to have the largest foraging force available at the time of the main nectar flow.

Expected nectar flows are very much at the mercy of weather conditions and accurate prediction is still a chancy matter in

Western Europe. It can, and does, happen that occasionally the nectar flow fails altogether.

Apart from heather, which is a special case, on the whole plants on chalky subsoils can be expected to yield nectar more freely than those on heavy clay soils.

There must be a constant supply of clean water. The collection of water is vital to the life of the colony and is sometimes overlooked when thinking of foraging.

These considerations inevitably lead to the conclusion that the beekeeper should not regard his own garden as the only possible place to keep bees. There are many advantages to be had from carefully selected out-apiaries and if beekeeping on any scale is contemplated, then out-apiaries more than two miles apart are essential. Without them many switching and queen-rearing operations are not possible. With farmers' increasing awareness of the usefulness of bees in agriculture it is not too difficult to find farm sites.

Apiary Sites

The consideration of suitable sites follows naturally from the previous section. In addition to the points already noted, the site should be readily approachable by a hard road, should have some shelter—a belt of trees on the prevailing windward side for instance and it should have some shade from noon-day summer sun. Frost pockets should be avoided. The apiary should not invite interference by cattle or vandals.

Hives

Whatever type of hive is decided on, they should all be of the same type for the sake of interchangeability of parts. Single-walled hives are more easily transported than the double-walled types. Single brood-chamber working is a convenience too, so it is as well to see that the hive you decide to use will be large enough to provide an adequate brood chamber for your particular strain of bee.

Bees

There is an ancient belief among beekeepers that bad-tempered bees get more honey than the sweet-tempered kind. I do not think this is so. They merely get interfered with less frequently, so are able to get on with their job of foraging. I am convinced that unnecessary and prolonged manipulations are responsible for a great deal of disruption of colony life. In any case, the beekeeper should want to enjoy his beekeeping and not have to be prepared for a fight every time he goes to his bees. Get rid of those incorrigibly spiteful bees.

There is a great deal in favour of using a local strain of bee which has become accustomed to local climatic conditions (perhaps 'inured' would be a better word sometimes!).

Have a regular routine of queen-raising and of culling undesirable or unsatisfactory queens. It is fairly well-established that queens do best in their second year.

Swarming
Decide in advance what you propose to do if you find any signs of incipient swarming. Inspections every nine days during May, June and July followed by application of your adopted system when necessary will prevent the loss of nearly all swarms and keep your colonies at top strength. I say 'nearly all' because however careful you are there is bound to be the odd sneaky one which will get away occasionally.

Generally
Every aspect of beekeeping from the beekeepers availability of free time through to local climatic conditions is so full of variables and imponderables that any attempt to lay down a hard and fast system of management which would be good in all kinds of situations and for all kinds of beekeepers would be merely foolish—and misleading at that. This is one of the charms of beekeeping. It is not a precise science. Each beekeeper will evolve his or her own method and will have personal preferences for different strains of bee, equipment and gadgetry. Long may it remain so.

To emphasize this point, if we take a quick look at the salient features of management adopted by Brother Adam in Devon, we can see that such a system, good for the harsh conditions on Dartmoor, based on his own strain of bee and with heather production always very much in mind, might be less suitable for other parts of the country with different strains of bees.

The System Used at Buckfast

After having been fed in autumn the bees are left alone until late February, the hives having been tied down to their stands with wire against the winter storms. In late February, as soon as weather permits, each colony is lifted on to a clean floor board.

In spring and as early as weather allows, strong colonies— that is those stronger than the average—are robbed of combs of brood and bees and these are given to weaker colonies. Most of the apiaries are about 40 colonies strong and, of course, combs are not transferred within apiaries, but are taken from colonies at a distance. This is 'equalizing'. The aim is to get all colonies at the same strength so that all can have the same subsequent treatment.

The main re-queening is done in spring. There is always a considerable reserve of young mated queens over-wintered in nuclei at mating stations on Dartmoor. The attempt is made to have two-thirds of all the honey producing colonies headed by second year queens. Young queens are taken from the nuclei, caged and given to colonies needing re-queening. The old queens from the colonies are similarly caged and given to the nuclei from which the young queens have been taken. They will remain there to keep the nuclei going until virgin queens are raised later in the year when they will be removed and replaced by the virgins. All queens are clipped before introduction.

Modified Dadant hives are used and during the part of the year when colonies are expanding they are confined to the number of frames they can cover. Fresh frames of comb or foundation are added as they are needed. There is no 'stimulative' spring feeding.

By mid-May and just before the brood chambers are fully occupied, the first supers are added. The extra room given is regarded as a help towards swarm prevention.

Regular weekly examinations are made from the end of March until the end of the swarming season. Any queen cells found are destroyed and fresh supers are added if this is found necessary at examinations. The honey crop is extracted as soon as possible after the end of the nectar flow.

Towards the end of July the hives are transferred to Dartmoor for the heather flow. Each is fitted with one super and each hive is secured against movement during transport by a simple yet ingenious device. Threaded steel rods pass through super and brood chamber and are screwed into brass plates let into the floor board.

At the heather additional supers are added if needed, but this is only done after careful consideration. Half-filled combs of heather honey are avoided if possible.

Only colonies that are very strong indeed are taken to the heather. After the ling has finished secreting nectar the supers are taken off using Porter escapes and they and the hives are brought back separately. Each colony is given a gallon (4.5 litres) of syrup on its return.

Because of the damp climate through ventilation is given for the winter. Strips of 1/8-inch (3mm) wood are laid over the frame lugs under the crown board and this allows the escape of moist warm air upwards via the two gaps created, with clean air coming through the entrance.

The above can only be regarded as a very sketchy outline of the management system adopted at Buckfast. There are many omissions and much has been telescoped, but I think enough has been said to emphasize that a regular method is used which will meet local conditions. Brother Adam is a charming and reticent man. He is a fine beekeeper of world wide renown—one of the few 'greats' of our times, however much he would object to my saying so. I am sure that he would be the first to admit that his system has been evolved over the years as a result of many trials and some disappointments. It is unlikely that it will remain forever unaltered. New discoveries and experiences will call for modifications. In our humble way we too must be prepared to adjust and alter our own methods.

3

Hives

In the wild, bees make their nests in any cavity which will satisfy the following criteria. It must provide shelter from rain and wind, have an internal ceiling to which combs can be attached, the cavity must be deep enough to accommodate the hanging combs and it must have an entrance small enough to be defended. All these requirements are met by hollows in tree trunks. They abound in nature. It seems to me that way back in prehistory bees were forest creatures.

Having once established a nest the colony would remain in it for many generations of queens (and even more of workers). European races of bees do not abscond from their homes unless impending disaster threatens; for example, starvation or the destruction of the nest site by high wind or forest fire. Bees are thought to have originated in India. In the warm climate there weather protection is not so necessary and races of bees have evolved which build their nests in the open. A few large combs will be built from branches of trees or in crevices in rocks. These races of bees do abscond, frequently and for no apparent reason.

Primitive Hives

The first use of bees by man was when honey hunting. Wild bees nests were found and robbed of slabs of honeycomb and grubs, both of which being consumed with relish as far as we know. A painful method of collecting one would have thought but one giving an almost complete diet in itself—carbohydrate from the honey and proteins and vitamins from the grubs and brood food. Not that primitive man gave much thought to balanced diet. He merely ate what he could get and what he liked. The next step was to sling hollow sections of tree trunks in the branches of trees by means of ropes and to try to induce swarms to enter them by means of 'baits' of small pieces of honeycomb. Then, presumably, someone got tired of always having to climb trees for his honey and started to keep hollow trunks on the ground where they are easier to manage. These were the first 'hives'. I sometimes wonder if the old adage should not be *'Laziness* is the mother of invention'.

Once the idea of static hives caught on they would be made of whatever materials were readily available locally. Indeed there was no other choice. In forest areas hollow tree trunks or large branches would be used as we have seen. Beekeepers in scrub country would make baskets of wicker—or perhaps domestic baskets made for other purposes would be pressed into service. The Greeks used wicker baskets with the mouth of the basket upwards. Across the open mouth wooden sticks would be laid for the bees to build comb from and some sort of cover, a stone slab or similar would be placed on top partly as a weight to keep the whole thing stable and partly as weather protection, although the arrangement was only possible because of the low rainfall. In agricultural regions baskets would be made of coiled straw rope sewn together with natural strands of material such as bramble stems etc. This is the familiar straw skep. The skep would be placed open end downwards on a stool. In arid mediterranean countries pipes of earthenware or clay would be used. When wood-working tools became available, sawn planks of timber would be lighter and more economical.

All these methods have been used for centuries and still are in undeveloped areas. Our modern hives are developments from the sawn plank variety. The great disadvantage with all these

Figure 6. Wild bees in a hollow log.

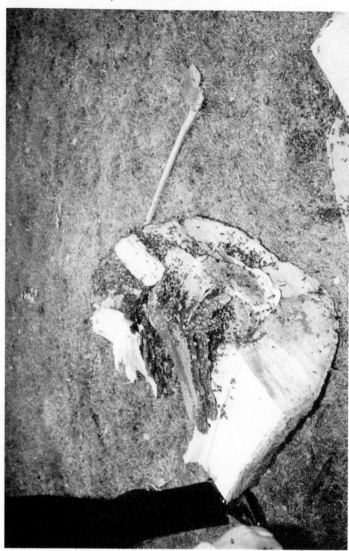

Figure 7.   The old trunk has been split with an axe to reveal part of the bee colony.

Figure 8.   A swarm in a skep.

hives was that the comb was fixed by the bees to the sides and ceiling of the hive. Management methods were difficult and some were impossible and the custom was to encourage those strains of bees which would throw swarms readily, secure these swarms for next year's stocks and to take the current year's honey by one of three methods.

1. From the more primitive hives pieces of honeycomb were removed from the hive, sometimes using a 'back door' if it was possible to make one.
2. By killing the bees. This was usually done by digging a shallow pit in which sulphur was burned. The hive was rested over this and the sulphur fumes killed the bees. Honeycomb could then be removed at leisure—and peacefully.
3. By driving the bees from the old hive into a new container. This was easily possible with straw skeps because of their light weight and small size.

In the latter method the skep to be cleared would be heavily smoked and inverted. Over the top would be fixed an empty skep of similar size by means of U-shaped iron spikes. The top skep would be held at an angle to the lower so that an opening would be left towards the beekeeper and, on the side away from him, the rims of the two skeps would be touching. Rhythmic drumming on the sides of the lower skep with the palms of the hands or short wooden staves would cause the bees to leave their combs and march up into the empty skep. Very few would take to the wing. The bees could then be united to a weak lot, left to build fresh comb if the operation was carried out early enough in the year, or returned to the original skep after some of the honeycomb had been removed. If performed by a competent beekeeper, this was quite a spectacular performance and was a popular item on the programme at Agricultural Shows and Honey Shows. It is said that this method derives from an inherited fear of the bees of instability of their nesting site. It is claimed that the rhythmic drumming puts the bees in mind of the regular tapping of branches in a situation where the nest is in danger from high winds. I think the story is apocryphal but it works and needs to be seen to be believed. At least it is less cruel and destructive than the other methods.

Modern Hives

In 1851 an American clergyman, L. L. Langstroth, noticed that bees would 'respect' a gap of about $\frac{3}{8}$-inch (9mm). If a space were less than say $\frac{1}{4}$-inch (6mm) they would close it with propolis and if it were more than $\frac{3}{8}$-inch (9mm) they would bridge the gap with brace comb. This simple discovery revolutionized beekeeping and is the basis of the construction of beehives and frames of all types. It meant hives could be made with removable combs and, provided this space (called a 'bee space') was carefully observed, the combs in their frames could be removed and replaced easily and without undue disturbance of the bees.

From a practical point of view the problem remains of which sort of hive is to be preferred. The thing to be borne firmly in mind is that beehives are designed for the convenience of bee-keepers—not for bees, who will be quite happy in any kind of box. 'Best' depends on what suits the beekeeper, the locality, the strain of bee and the kind of management procedures adopted. It is curious how fiercely beekeepers hold to their pet theories of hive size and design.

There have been one or two attempts to fabricate hives from man-made materials—mostly petro-chemical derivatives—with, it seems to me, very limited success. There was even one monstrosity composed of aluminium angle pieces and glass, now mercifully discontinued. Hives made from timber seem to be far and away the best at the moment. Although good timber is not cheap, and anything but good timber is useless, a hive, well constructed from matured timber will last for many many years and will require little attention other than the need for weather-proofing. Some years ago I bought an old C.D.B. hive merely because I like the appearance and I wanted an old hive for demonstration purposes. The previous owner had allowed the hive to stand on the ground directly and the lower ends of the legs had rotted. These I renewed, but apart from this the hive was, and still is, as sound as a bell and perfectly usable. No parts of the hive body have ever been renewed. This type of hive was popular in Ireland around the turn of the century and as far as I know has not been produced commercially for many years. As far as I can discover, my example is at least fifty years old and may be as much as seventy to eighty years old. This is the kind of longevity and lack of deterioration one can expect from a well-made timber hive.

I also have a number of hives made to the original National specification during the war years 1939-1945. These too are sound and the bees seem happy in them. Of course this is not an isolated instance. There are many old hives in current use which outlive their beekeeping owners and continue to give good service. So, although the initial cost of a new hive may appear high, it is a good bargain in terms of the period of use which may reasonably be expected.

Making Your Own Hive

The home construction of hives is a matter of simple carpentry and well within the capabilities of the average man (or woman). It is absolutely vital, however, to keep strictly to standard dimensions in order to maintain interchangeability of parts. Working drawings of hives can be had from Beekeeping Associations and the Ministry of Agriculture pamphlets on hives are extremely helpful, although not sufficiently detailed. Western red cedar is often recommended as suitable for hive construction. It is true that it is light, contains its own preservative and is easy to work. However good quality red cedar is hard to come by and expensive. It is too soft and corners of hive bodies get damaged through repeated use of the hive tool. Good quality deal is a sounder investment in my view but it is absolutely essential that it is sufficiently matured and quite dry. The disadvantage of the greater weight as compared with cedar is slight. Most hives offered by commercial bee appliance manufacturers will be made of red cedar.

Single- vs Double-walled Hives

Two types of hive are available—double- or single-walled. One variety of the former and four of the latter are most commonly found at present and we will confine our attention to these. From time to time fresh designs are thought up to satisfy pet theories. They enjoy a limited popularity among a few devotees and are perhaps best left to the man who wants to experiment. This is not to say that they are useless, far from it.

I do not think that any aspect of beekeeping has given rise to more acrimony between their adherents than the double- versus single-walled hive debate; so I will not add to it other than to say that in the milder, damper conditions prevailing in the south-west of Britain the double-walled hive is extremely popular and bees do well in it. The arguments for and against fall into four parts:

1. *Temperature within the hive.* It is claimed that a double hive is warmer in winter and cooler in summer because of the jacket of air surrounding the inner boxes. It is true that when there is a rapid rise or fall in outside temperature there is a time lag of maybe 30 to 60 minutes before the temperature inside the hive equalizes. Extensive research with thermocouples has shown that in winter the temperature as close as four inches (10cm) to the hive wall was seldom more than 1° to 2°F above that of the outside air. There is no reliable evidence that the type of hive selected is of any real importance as far as temperature is concerned.

2. *Resistance to wind and rain.* Here again there seems little to choose between the two types providing that both are equally well made. Many double hives have gable shaped roofs. In high wind these blow off rather more easily and a flat roof is useful to stand things on!

3. *Ventilation.* Excessive moisture within the hive, especially during the winter months, is probably one of the greatest enemies of successful beekeeping. Bees can stand a fair degree of cold if they are in a sound dry hive and have adequate food. It does seem that very few cases of mouldy comb are found in double hives. The inner boxes are usually made of light timber and, at least in early models, there were always gaps where the inner boxes fitted over the frame lugs of lower boxes.

   The inner boxes are never painted or treated with pre-servative. However it may be that the designer, in making what he thought to be a warmer hive, has created one which is better ventilated. Users of single hives get over the problem by one or a combination of some of the following:

—Providing a large bottom entrance covered with a strip of zinc perforated with $\frac{1}{4}$-inch (6mm) holes to keep mice out.

—Leaving the feed hole in the crown board open or leaving the crown board off altogether, or even raising the crown board slightly on four matchsticks.

—Placing an empty super on the crown board to make an air space.

—Having no packing in the empty super or packing of a porous nature, e.g. clean dry sacking.

—Making a top entrance as well as the usual lower one. This is not very popular in the United Kingdom, but is common in the United States.

These ventilation devices are needed only in winter. In summer the bees' own air conditioning will keep the hive aired and sweet.

4. *Ease of manipulation and transport.* Double hives are certainly more bother during manipulation even if only because there are more parts to deal with. Often the lifts have to be put on the right way round as they are not always square.

There can be no argument that, if hives have to be moved, the single variety is much easier to make secure and bee-tight, to load onto a truck or into the back of a car, and its simple box shape is more economical with space.

My own conclusion is that there is very little difference between the two types as far as the health of the bees is concerned. Double hives are more expensive but their appearance is more attractive in a garden setting. The boxy shape of the single hive is hardly a thing of beauty although it may well be an eternal joy to its happy owner.

The choice must lie with the beekeeper. If he proposes to keep only a few hives in his garden then he may prefer the look of white painted double hives. If he intends his beekeeping to be more extensive and migratory then the single variety will be his choice.

The only example of the double-walled hive remaining in popular use is the W.B.C., so-called after the man who

Figure 9.   Older type of C.D.B. hive, made circa 1910. The lower box is a single-walled brood chamber. The upper box overhangs slightly (but is bee-tight) to accommodate an internal super.

introduced it, Mr William Broughton Carr. At one time there is no doubt that this was the most widely-used type of hive in this country and it is still liked by a good many beekeepers. Its popularity has now been overtaken by the simpler single-walled hives.

It consists of a series of outer cases or 'lifts' surrounding inner boxes, which hold ten British Standard frames, so that there is an air space all round and above the inner boxes. No doubt the idea was inspired by the construction of cavity walls in houses and the superior insulation conferred by this design. The lifts and inner boxes rest on a floor which extends to an alighting board and a shallow porch holds sliding wooden strips so that the width of the entrance can be infinitely varied. Slightly splayed legs raise the floor above the ground.

It is sometimes dismissed by experienced beekeepers as 'inefficient and amateurish'. I think this is perhaps a little unfair and smacking of professional snobbery. Efficiency and the lack of it lie with the beekeeper himself and what is wrong with being an amateur anyway? Although I prefer, and use, single-walled hives myself I know a good many beekeepers who use W.B.C. hives very efficiently.

The single-walled hives most readily available are the Modified National, Modified Commercial, Smith, Langstroth and Modified Dadant. Principal dimensions of these are given in Table 2.

Which Single-walled Hive?

All these hives have their own (sometimes vociferous) devotees. We have already looked briefly at the double versus single wall controversy and supposing this is settled and the single type decided upon, then the question is which specific design? From the construction point of view no special difficulties are likely to arise. Some are easier than others, but all are well within the capabilities of the normal beekeeper (if there is such a person). All the types listed are stock items at the large bee appliance manufacturers and in each case shallow boxes or supers are made for the storage of honey. These shallow boxes have precisely the same plan dimensions as the deep boxes intended as brood chambers but are about two thirds the depth.

The two most important considerations in choosing the most suitable design are i) the available comb area in the brood chamber and ii) whether the bee space is provided at the top or bottom of the boxes.

i) *Size of Brood Chamber*
All hives consist in essence of open-ended boxes which can be

Table 2.   Hive Dimensions

Hive	Brood Chamber				Frames (in inches)				Number of frames in brood chamber
	external dimensions (in inches)			Bee space T = top B = bottom	Top bar	Frame length	Frame depth	Approximate area available for comb building—both sides of one frame (in square inches)	
	wide	long	deep						
National	$18\frac{1}{8}$	$18\frac{1}{8}$	$8\frac{7}{8}$	B	17	14	$8\frac{1}{2}$	185	11
Smith	$16\frac{3}{8}$	$18\frac{1}{4}$	$8\frac{7}{8}$	T	$15\frac{1}{2}$	14	$8\frac{1}{2}$	185	12*
Modified Commercial	$18\frac{5}{16}$	$18\frac{5}{16}$	$10\frac{1}{2}$	B	$17\frac{1}{4}$	16	10	260	12*
Langstroth	$16\frac{1}{4}$	20	$9\frac{9}{16}$	T	19	$17\frac{5}{8}$	$9\frac{1}{8}$	255	10
Modified Dadant	$18\frac{1}{2}$	20	$11\frac{3}{4}$	T	19	$17\frac{5}{8}$	$11\frac{1}{4}$	320	11
	internal dimensions								
W.B.C.	$14\frac{1}{2}$	$15\frac{3}{8}$	$8\frac{7}{8}$	B	17	14	$8\frac{1}{2}$	185	10

*Although it is possible to get 12 frames into Smith and Modified Commercial brood chambers, this is a tight fit and leaves no room for a dummy board or for manoeuvre.

placed one upon another to increase the capacity of the space made available for brood-rearing, or for the storage of honey. When properly made, the edges of the boxes marry and are bee-tight. Many beekeepers like to have the brood area confined to one box and find that ten or eleven British Standard brood frames are hardly enough for their strain of bee. Congestion results and this leads to inefficient management and increases the risk of swarming. These people will prefer a more capacious brood chamber than the single National, for instance, the Modified Commercial. This is almost compatible with the National, its $18\frac{5}{16}$ inches (46.5cm) square plan being very close to the $18\frac{1}{8}$ inches (46cm) plan of the National. In fact National deep and shallow boxes can be used as honey supers over a Modified Commercial brood box.

There are more National type hives in use in the United Kingdom than any other. The original National had double walls on opposite sides in order to provide a kind of shelf on which the lugs of the frames could rest. Later designs have single walls on all sides and accommodation for the frame lugs is provided by box-like extensions on two sides. These also make very secure hand holds. All parts are interchangeable and the brood nest can be expanded by allowing the queen to run on more than one box—either two deep or one deep and one shallow.

The Smith hive is another very simple hive to make. It consists simply of a rectangular box of the dimensions given in the Table 2 with rebates cut into the inside top edges of two opposite sides. The frame lugs rest in these rebates. The hive takes British Standard frames with short ($\frac{3}{4}$-inch) lugs instead of the normal long ($1\frac{1}{2}$-inch) lugs.

As has been noted in the table, it is possible to get twelve Hoffman self-spacing frames in a Smith brood chamber—but only just. No room is left to prise apart adjoining frames and most people prefer to use eleven frames plus a 'dummy' frame. This is a piece of $\frac{3}{8}$-inch (9mm) board 14 inches (35.5cm) by $8\frac{1}{2}$ inches (21.5cm) fastened to a top bar. Staples or screw eyes are driven in near each corner on one side so that they protrude about $\frac{1}{4}$ inch (6mm). This side faces the hive wall. On the other side of the dummy board slips of wood $2\frac{1}{2}$ inches (63mm) by $\frac{3}{8}$

inch (9mm) by ⅜ inch (9mm) are fixed vertically so that they will butt against the self-spacing shoulders of the first frame. Removing such a dummy will give working space to get the first frame out when manipulating. Metal ends (see under 'Frame Spacing', p. 54) are not suitable for use in Smith hives.

The two remaining hives are of American origin. They are in practically universal use in the United States, especially the Langstroth which has also become fairly popular here. Both give more comb space than the British National and both are of simple design.

ii) Top or Bottom Bee Space?

This is the second important consideration in choosing a hive. When the space is at the top of the box i.e. above the frames, handling becomes much easier. Supers or second brood chambers can be taken off and slid on without being scraped across the top bars of the lower box as is the case when the bee space is at the bottom of the box and consequently the top bars of the frames are level with the top edge of the box. It is claimed too that bees build less brace comb between the tops of the lower frames and the bottom bars of the upper frames. I find it difficult to see why this should be. If the bee space is correct, it would not seem to matter very much where it is.

Preservation of Hives

Although it is true that a well-made hive will give good service for many years, it will stand in the open in all seasons and will have to withstand the onslaught of extremes of weather, so it is only prudent to give some thought to the question of regular maintenance and weather-proofing.

The outer lifts of W.B.C. hives and the outsides of single-walled hives are best treated as would be any other timber intended to remain in the open. Cedar wood contains its own preservative and needs no further treatment, but an annual coat of linseed oil will be an aid to weather-proofing and improves appearance. If construction is of deal or any other timber it can be either painted or treated with wood preservative. If painted, the normal procedure should be carried out. Any knots should be treated with knotting followed by all-over coats of primer and undercoat and then two coats of top coat, rubbing down between each application. White, cream or pale green top coats give an attractive finish. In out-apiaries a darker green finish helps to conceal the hives from the unwelcome attentions of two-legged predators. In W.B.C. hives pay particular attention to the 'soles' of the legs. Being the end grain of the timber they are most liable to soak up damp from whatever they are standing on. It helps if a small square of builders damp course is tacked to the bottom of the legs after painting—a double layer is even better. Inside W.B.C. boxes are usually left unpainted.

As an alternative to painting the use of a proprietary wood preservative is quite satisfactory and quicker to apply, but great care must be taken to see that the preservative contains no insecticide or fungicide. Some do, and suppliers are usually able to say which they are.

Linseed oil can be used too as can creosote, but in this case hives so treated *must* be allowed to air in the open for at least three months and longer if possible. One disadvantage of using wood preservatives is that the joints at the corners are not sealed.

In fairness it must be reported that some beekeepers hold that waterproof paint should not be used on the outside of hives which should be treated in some alternative way so that the wood can breathe. This is not the place to enter into a long and tedious argument on the point, but I would merely say that almost the first thing bees will do on their side of a new hive (the inside) is to coat it all over with a propolis/wax varnish. A quick look at the inside of any hive that has been in use for some time will confirm this.

All nails and screws used in construction and repair work should be either sheradized or galvanized in order to avoid rusting.

Frame Spacing

In the natural state bees build their brood combs in such a fashion that two bees can pass back to back between them. This works out at $1\frac{3}{8}$ inches (35mm) from centre to centre. Combs for honey storage are a little closer. The bees seem content with space enough for one bee only.

In movable comb hives therefore, it is wise to space your combs at similar distances and over the years different ways of doing this have been evolved.

Self-spacing Hoffman Frames

These have side bars, the top two inches (50mm) or so of which are widened into 'shoulders' $1\frac{3}{8}$ inches (35mm) across. One edge is left plain and the other chamfered to a sharp edge. In use the chamfered edge of one frame bears on the plain edge of its neighbour (see Figure 10a). The idea of the chamfer is to limit the area of contact and so reduce the possible amount of propolizing. The shoulders of Hoffman frames maintain correct spacing and also hold the frames firmly in a vertical position. A light smear of grease will discourage propolizing which could widen the spacing if left uncleared. By pressing the frames tightly together this tendency is reduced. It is possible to buy plastic clips which will convert British Standard frames with $\frac{7}{8}$ inch (22mm) side bars into Hoffman type self-spacers.

Self-spacing with Screw Eyes

The same effect as with Hoffman frames can be obtained with the use of screw eyes. 'Medium No. 4' screw eyes have a ring of about $\frac{1}{2}$ inch (13mm) and are about right. Care must be taken in fixing them. Hold a frame with the top bar uppermost and pointing away from you. Insert one screw eye into the right side of the nearest side bar just below the top bar. Screw right in until the ring meets the wood and is horizontal, and then insert a second screw eye 2 inches (50mm lower). Turn the frame through 180° and repeat on the other side bar. Place a screw eye in each side wall of the hive to meet the nearest frame on the side where it has no screw eye. This is a good and simple method of spacing. The points of contact are extremely small and by screwing the eyes in or out a half-turn or so, adjustments can easily be made to get the spacing exactly right.

One screw eye in each side bar will look after spacing, but by using two, a vertical hold is achieved. It is surprising how frames will swing when transported, for example in the boot of a car, sometimes to the detriment of the unfortunate bees.

Metal Ends

These are stamped from sheet tin and slip over the lugs of the frames, allegedly to give spacing (see Figure 10b). They bend very easily and bees seem compelled to smother them with propolis so that spacing is problematical and they give no vertical hold. If used on super frames they have to be removed before the frames are loaded into an extractor. In my view they are quite horrible and the best place for them is the dustbin.

Yorkshire Spacers

These are also stamped from sheet tin and are placed, two to each side bar, one just below the top bar and the second two inches below (see Figure 10c). They are effective but have a rather large area of contact. Propolizing tends to enlarge the spacing.

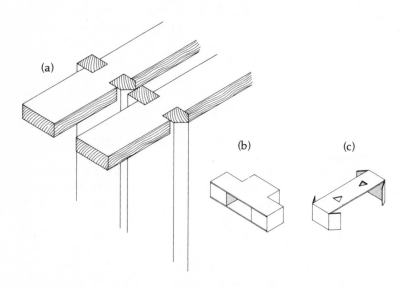

Figure 10. (a) Hoffman self-spacing frames. (b) Metal spacer. (c) Yorkshire spacer.

The four methods outlined above apply to brood frames.

Castellated Runners

These are strips of metal fixed inside the side walls of supers in place of the usual runners. They have slots cut in the top edge in which the lugs of the super frames rest.

Panel Pins

Frames are carefully spaced in the supers and 1-inch (25mm) panel pins are driven in for about half their length on either side of each lug. Frames can be lifted out vertically for extracting and subsequently returned without, of course, the need to remove any metal ends etc. Little propolization occurs, perhaps because the bees have access all round the lugs.

4

Handling Bees

Half the pleasure in keeping bees is to be gained from the smooth and efficient handling of stocks with a minimum of disturbance and, after an examination, leaving the bees working in much the same way as they were before you started. Half the people who give up beekeeping do so as a result of a disastrous experience with a thoroughly alerted and aggressive stock. The trick is, of course, not to let a stock get into this angry state. Work with your bees not against them. If you find that a usually quiet stock 'boils up' when you open up the hive, this may quite likely be due to some condition within the hive. Discretion is the better part of valour so close up quickly and smoothly and leave them alone for a few days. Try again and you will probably find that the mood has changed completely and they are their usual well-behaved selves. You will wonder what set them off in the first place.

I am sure that the best way of starting to learn to handle bees —and this is not a difficult art to acquire—is to take every opportunity of watching a real bee master at work and to assist if possible by acting as his 'smoker boy'. It will seem at first as if he has some kind of magic touch which enables him to handle bees and to move frames about, even to scoop up handfuls of bees without getting stung and without causing any kind of uproar. It is not true of course, neither is the widely held belief (by non-beekeepers) that bees know their owner. Watch carefully and you will see that everything he does, he does gently. He moves slowly and deliberately and always has whatever equipment he needs close to hand. He clearly likes his bees and is familiar with their ways so that he can interpret their moods and anticipate their reactions to conditions. It has probably taken him a lifetime to reach this blissful state, so I do not suggest that you will become like him overnight. But I do suggest that you can achieve some of this expertise with a little thought and practice fairly quickly.

Beginners, remembering all they have been taught and all they have observed about handling bees, will often do quite well in a short time. After a while they may become a little over-confident and bang a stock about or open a hive in unsuitable weather. They get a sharp reminder from the bees to be more careful in future. As in other walks of life, it pays not to take liberties!

If the following points are borne in mind the road to success will be that much quicker and more pleasant.

A bee cannot withdraw her sting after thrusting it into the resilient flesh of a warm-blooded animal, so she will live only a few hours after stinging. She will not use her sting except under provocation or in panic. Her hereditary enemies are marauding animals and birds and forest fires. She has two powerful instincts: first she has an impulse to collect nectar whenever it is available, and second to protect the colony, especially when young brood is present. Most swarms can be handled with ease. This may be due to two factors. The bees will have gorged themselves on honey before issuing in search of a new home. Unless the swarm has been hanging out for some time, the bees will still be full of food. Also there will be no young brood to defend.

Because of efficient communication and scent a colony will, in a very short time behave in accordance with the reactions of individual bees composing it. Annoy a few bees and the colony will quickly become alerted and aggressive. Frighten ten per cent into gorging honey and soon many more will follow suit.

Things which Lead to Contentment
— Presence of brood, a laying queen and plenty of stores with a few open cells.
— Quiet deliberate movements.
— Unobstructed flight path. (Do not stand in front of the hive or allow others to do so.)
— Warm still weather.
— A nectar flow.
— Keeping the combs covered.
— Clean smooth, light-coloured covering cloths and clothing. A white cotton boiler suit is ideal. Some bee-keepers find that nylon clothing annoys. I think this may well be so if movement and rubbing induces an electric charge on the nylon. Bees in flight and on the combs (wax is a poor conductor) carry an electrical charge and a nearby large mass with a high potential might well reverse the polarity on the bees. 'Touchiness' in bees at the approach of a thunderstorm is well known.

Things which Frighten
— Smoke—best if cool and aromatic. Perhaps that is due to an inherited fear of forest fire.
— Rhythmic drumming on the outside of the hive.

Things which Subdue and Control
— Proper, not excessive, use of smoke. Too much smoke will frighten the bees; it will not stupefy them as is sometimes imagined.
— A fine spray of water.
— Sprinkling with weak sugar solution. This is particularly useful when stores are a bit low and when no open cells are handy. The bees seem so concerned about cleaning themselves up and licking up the drops of syrup from the top bars that they tend to forget to have a go at you.

Things which Annoy
— Sudden movements across the tops of exposed frames.

— Jarring and the jerk when a stuck frame is prised loose.
— High winds and cold weather.
— Thundery weather.
— Human breath (do not breathe into the hive).
— Perspiration.
— Hairy clothes, especially if brown. Blue is not a good colour either.
— Crushing bees.
— The smell of their own venom. All beekeepers try to avoid being stung if only because a bee which has used her sting is a dead bee in a few hours. If you do happen to collect a sting, and this happens to the best of beekeepers occasionally, try not to jerk but immediately blow smoke over the site of the sting which should be removed. The edge of the nozzle of the smoker is a useful tool for this. If the nozzle is too hot then you will know that you have been smoking too enthusiastically.

Opening the Hive

With these points in mind, when you prepare to do your first solo manipulation choose a day when it is calm and warm. Mid-afternoon is best. If the bees seem to be coming home loaded with nectar, so much the better. Bees heavy with nectar have a characteristic stern down and legs forward flight and are always in a great hurry to get back inside the hive. Wear a clean white boiler suit or a light-coloured linen jacket past its best days, a veil, trouser legs tucked into wellingtons and gloves. As your skill and confidence grows you will most likely dispense with gloves—but always have them handy, just in case. For reasons which we need not go into, but which are obvious, trousers are best for ladies.

Make sure your smoker is well alight. Rolls of old sacking or corrugated paper will do, so will *dry* rotted wood. Make sure that the corrugated paper has not been fireproofed. Some packing materials are treated in this way and it is disconcerting to find that when you are all prepared, the confounded smoker fuel is fireproof and will not ignite. Light your roll of fuel and when it is well smouldering place it in the fire box of your smoker lighted end downwards.

Have spare smoker fuel and a box of matches in your pocket and your hive tool, cover cloths and a spare brood chamber, if you have one, ready to hand. I like to have a well-washed washing up liquid container filled with weak sugar solution (1 lb/450g sugar to 1 pint/550ml of water) by me.

Make a Plan

There is really no point in opening a hive unless you have in your mind a clear idea of what you want to look for and what you intend to do. Stand behind the hive—right at the back if you keep frames parallel to the entrance and by the rear left hand corner of the hive if your frames run at right angles to the entrance. From now on try not to move your feet more than is absolutely essential. Smoke the entrance with three or four medium puffs. Wait two minutes. The smoke will frighten the bees and make them rush to open cells and fill their honey sacs with honey. The maximum effect of smoke occurs in two or three minutes. During this time you can be quietly taking off the roof and any packing materials on the inner cover. Lay the roof down gently on your left. Do not bang it down. Bees are sensitive to vibration.

Remove the supers or inner cover by inserting your hive tool at a corner between the supers and the queen excluder or between the inner cover and the brood chamber, as the case may be. Lever upwards, gently puffing a *little* smoke along the crack you have made. Make sure that no frames are stuck and are being lifted. If there are any stuck frames, free them with your hive tool.

Once you start lifting, pause if you have to, but do not lower. If you have to change your grip or wait for any other reason, keep the boxes apart with your hive tool or a piece of wood until you are ready to carry on. Bees will be running about on the space over the frames you have created and the last thing you will want to do is to crush any of them. If you are removing only the inner cover, examine the under side carefully, holding the cover over the open hive as you do so. Make quite sure the queen is not there. If she is, let her run back into the hive; put her there if she will not run in the direction you want. The object of holding the cover over the hive is to ensure that if the queen is on it and falls off while you are moving it about she will fall into the hive and not into the long grass round the hive you meant to cut last week-end but forgot to.

If you are taking off supers, remove the queen excluder in the same way, again looking carefully for the queen on the underside. Cover the frames with one of your cover cloths. These are simply lengths of clean white cotton cloth (cut from an old sheet perhaps) about two inches (5cm) or so bigger all round than the area of your hive top tacked to a piece of dowel rod at each end. You will need two of these. Their function is to keep direct light from shining down into the brood chamber and to keep the bees down there. The cloth should be put on with the dowel parallel to the frames.

Starting at one end roll the cloth back on the dowel until the first frame is exposed. Free this carefully with your hive tool, lift slowly and gently by the lugs and hang it in the spare brood box. If you have no spare box place it on the ground at the side and to the front of the hive. Before putting the frame down look again for the queen, although it is unlikely that she will be on an end frame. You will find these mostly used for stores. When lifting frames, take care not to roll the bees against the next frame. This never fails to annoy them.

Inspecting the Frames

You are now in a position to go through the brood box and examine every frame. By removing the outside frame you have made a little working space for yourself. With subsequent frames you will be able to lever them back into the space before lifting out, then after examination replace them towards the examined frames. In this way the space you have made travels across the hive as you work. You need only expose one frame at a time if you roll back cover-cloth number one to clear the next frame and unroll cover-cloth number two to cover the frame you have just replaced. This will prevent the bees boiling up between the frames. All your movements should be slow and deliberate. Although you have got most of the brood chamber covered, still do not move your hands across the tops of the frames. Try to get into the habit of moving your hands round the outside of the hive up to the frame you want to lift out and then do so by gripping the lugs between the thumb and forefinger.

Bees will propolize frames to runners and the neighbouring frames and metal ends to other metal ends, especially if carpentry has not been too accurate. Use your hive tool to prise them free—but gently. That wrenching jerk ending with the frame jumping clear is just the thing to alert the bees. Never turn a comb so that it is horizontal. Always hold it over the hive so that should the queen drop off she will land back in the hive. The natural way to look at a frame is to hold it up by the lugs with the top bar horizontal and the frame and comb hanging down vertically. To see the other side lower one hand and raise the other. Turn the frame through 180°.

In this way you will be able to scan both sides of the frame and will never have had the comb horizontal. In warm weather a comb half-full of stores can easily sag if held horizontally and new nectar and half-processed honey will pour out.

Use smoke sparingly and puff it across the tops of the frames when it is needed, never down between the frames. Carry out your examination smoothly and without undue haste, but do not prolong it a moment more than is necessary. It is not even necessary always to take out every frame. For instance, if you feel that you need to make sure that the colony is queen-right (a colony is said to be queen-right when it is headed by a mated and fertile queen), you do not need to actually see the queen. As soon as you see eggs or very young larvae you can close up at once. You have satisfied that the queen is there and is laying.

Any odd bits of wax and brace comb you scrape off the top bars should be placed in an old tin or some other receptacle.

Tackling a 'Touchy' Colony

The time may come, not in your first years of beekeeping I hope, when you may have to tackle a bad tempered colony or a 'touchy' one in unsuitable weather. Two possible ways of reducing the unpleasantness are:

1. *Move the hive bodily some yards away.* You will need another pair of hands to do this. Place an empty brood chamber on the original site of the hive with some kind of

cover on it. Go back to the moved colony and do whatever has to be done. The flying bees—those most likely to be troublesome—will return to the old site and cluster in the empty brood chamber so you will have fewer bees to contend with.

2. *Anaesthetize the bees.* This can be done by using in your smoker, fuel which has been spread with a solution of nitre and dried. Insert the *unlighted* end of the fuel into the fire box this time otherwise you will have a firework in your hand. Give the colony a good smoking. The nitrous oxide emitted will temporarily knock the bees out, but work quickly. When they come round the bees will be worse tempered than ever.

5

Behaviour

I do not think it is possible to keep bees successfully without understanding a little about their behaviour in their environment, the senses by which they become aware of their surroundings and the responses which are elicited. To do this let us glance briefly at the various senses of the bee, which differ in some respects from ours, and the effects of stimulating these senses both on the individual bee and, more importantly, on the bee colony as a whole. Bees are social animals and the behaviour of the colony is more than the sum of the behaviour of all the individuals comprising it. The beekeeper rapidly discovers that the viable unit is the colony and the individual bees are the constituent parts. They work for the good of the whole colony and cannot live for very long away from it.

Vision
The facets of the compound eye (about 6,300 in the worker, 3,900 in the queen, 13,000 in the drone) project dots of varying intensity of light on the retina and so build up a kind of mosaic picture. Perception of detail is not therefore very acute. It has been estimated at not more than one hundredth of that of man.

There is no focusing mechanism. However, to make up for these apparent deficiencies the bee has a very good perception of movement and flicker. She can detect flicker frequencies of about 54 per second. Because of these characteristics the attention of bees is more readily drawn to broken patterns and if these are moving gently, for example in a gentle breeze, then the attraction is even greater. Flowers with narrow radiating petals or with a broken pattern of petal arrangement are likely to be more attractive than those consisting of large overlapping petals giving the effect of a solid block of colour are less so.

The range of colour vision is rather different from ours too. Bees can see more into the ultra-violet end of the spectrum than we can but not so far into the red end. For instance white flowers reflecting ultra-violet light appear exceptionally brilliant to bees while a large red flower such as a rose, attractive to us, will seem to them to be a large black blob devoid of colour or shape. About 30 per cent of conspicuous flowers reflect ultra-violet light; for instance evening primrose and red poppies reflect ultra-violet light well. Bees can also detect the plane of polarization of light, which we cannot, and use this ability in navigation.

The three ocelli, or simple eyes, on the top of the head certainly react to light, but are incapable of forming an image. No good evidence is available as to their functions (if any). They may be vestigial or, as has been suggested, useful in telling the bee which is 'right way up'. Owing to its light weight it may be that the bee has a poor sense of gravity and, as far as I know, it has not yet been shown that there is any organ of balance comparable to the cochlea in the mammalian ear.

Taste and Smell

The senses of taste and smell are separate in man, although closely allied, but in the bee the two seem to merge in some instances. Bees can perceive odours from distant sources but can also perceive them by direct contact with sense organs on the antennae. The antennae can perceive the scent of sugar, water and floral perfumes at a distance and there is reason to believe that this detection is directional—in a sense rather like binocular vision. Degrees of sweetness can be detected and bees will always prefer to work sources of nectar with high sugar concentrations rather than lower. They can also distinguish between different kinds of sugar and show a marked preference to those of high food value—sucrose, glucose, fructose, maltose, trehalose, melizitose—precisely those pleasant to man. The first three sugars are the most frequently appearing in plant nectars. The others do occur, but less frequently. Rather curiously, octoacetyl sucrose, which is intensely bitter to man, in fact uneatable, seems to be quite tasteless to bees and for this reason has been used to 'denature' sugar to be made available to beekeepers for feeding their bees (if they are foolish enough to do so) but is unusable for human consumption. Although the antennae are the seat of most of the organs of taste and smell some occur on the front legs and possibly the feet too.

Hearing
Bees do not hear as we do nor do they possess 'ears' analagous to those of vertebrates, but they are sensitive to vibration. They detect the vibrations of sounds through groups of elastic fibres in the tibia of their legs. These organs are very sensitive to mechanical vibration which, for this purpose might be regarded as a very low sound. They are so different in construction from the membraneous ear drum common in mammals that the range of tones and variations in pitch must be totally different.

Electricity and Magnetism
These two phenomena are all-pervading throughout the universe in which we live and it is curious that we know so little about their fundamental nature. What we *do* know is a little about the effects they exert on the environment. In recent years biologists have been considering the effects of magnetism on living organisms and have discovered that it is much more far-reaching than had been supposed. This book is not the place for an exhaustive analysis of the subject, even if I were capable of doing this—which I am not—but a brief look at how practical beekeeping might be affected is not out of place.

Electro-magnetic charges can be present on the surface of most objects especially those made of materials which are poor conductors. Most biological material is a poor conductor. Such charges are surrounded by magnetic fields and when a charged object moves in a magnetic field an electric current flows. Since bees are electrically charged and they move through a charged environment they must be subject to many electro-magnetic forces. They will be affected by mineral deposits (especially iron) below the surface of the earth, such things as electric power lines above it and the approach of thunderstorms and other electrical disturbances due to weather conditions. The effect of thundery conditions has been known for many years and beneath power cables is usually a bad place to position a hive, although bees bred in the vicinity of high tension cables have been known to do well.

Most of a bee's body has a charge of low potential, but membranes and glandular surfaces carry high charges. Antennae usually carry opposite charges and the polarity can be quickly reversed, apparently at will, one antenna becoming negative where it was positive before and the negative antenna changing to positive—all within the space of one second. How can these electro-magnetic effects influence bees? Little is known at present but research is proceeding. However it is probable that:

— Low frequency electric fields increase the bees' metabolic rate.
— High frequency fields are likely to lead to bees absconding. This has been observed.
— Electric charges may be one of the means of colony identification.
— Electric shocks make bees aggressive so it would seem prudent for a beekeeper not to wear clothes made of synthetic fibres which generate static electricity when rubbed. Cotton garments seem much better.
— It is well known that bees dislike sweaty people and animals. This may not be due only to odour but in part to the fact that the moisture on the surface of the sweater increases electrical conductivity.
— There is some reason to believe that bees prefer to build comb along lines of the earth's magnetic field unless otherwise prevented.

It is certain that electricity and magnetism have a powerful effect on bees. Much remains obscure and further research may well yield useful and practical results for the beekeeper.

Clustering

In social animals mutual attraction is a frequently-observed phenomenon and bees are no exception. The habit has survival value, especially in the case of bees. If a group of, say, one hundred bees are queenless, they will always tend to congregate in clumps. This can often be seen during a long manipulation of a colony during which the queen and different groups of bees become separated.

In a queen-right colony activity will centre round the queen, wherever she is.

The most obvious example of clustering is, of course, the action of a swarm when settling. The swarm has *not* lost its queen, but conditions are not normal and departure from the norm too will result in clustering. The beekeepers task of taking a swarm would be much more difficult, if not impossible, without this habit of clinging together.

Another example is wax production. When bees are at the right stage of maturity, the colony needs more comb and there is plenty of honey available, either in the hive or in the case of a swarm carried in thousands of honey sacs, the wax producers gorge with honey and hang motionless in festoons. Each clings by her front legs to the hind legs of the bee above. In twenty-four hours tiny wax scales are exuded from the wax plates on the abdomen. The wax scales are passed via the middle legs and front legs to the mandibles where they are kneaded with a frothy substance coming from the mouth parts. The wax becomes malleable and is available for comb building or repairs. This is just one more example of the many instances of the mixing of raw materials with glandular secretions which is so much a part of the bees life.

Food Exchange and Nectar Processing

A great deal of food exchange takes place within the hive and it is likely that this is the foundation of the social life of bees. With the food exchanged, minute traces of hormones, pheromones and other chemical substances are passed from individual to individual throughout the colony. The process is rapid. Six marked bees were fed with a small quantity (about 20ml) of sugar syrup to which had been added a trace of radioactive phosphorus. The six bees collected the syrup in $3\frac{1}{4}$ hours, carrying it back to their hive. The colony was small—about 25,000 bees in all—housed in one National brood chamber with two supers. Bees were tested for radioactivity after five and twenty-nine hours. Results were as follows:

Table 3. Results of Radioactive Food Exchange Test

Bees	After 5 hours percentage radioactive	After 29 hours percentage radioactive
Foragers	62	76
from brood chamber	18	43
from upper super	21	60
from lower super	16	53

After two days 85 older larvae were examined. All were radioactive. This was only one of many similar experiments which have confirmed rapid food exchange.

Larvae are fed by many nurse bees in turn who do not seem to look after any particular groups of larvae, indeed they seldom return to a larva for a second time. In this way even the juvenile forms participate in the dissemination of minute traces of important substances.

I do not think this process consists of hungry bees soliciting food from others—except in the case of drones. It seems to originate in the processing of nectar into honey. On returning to the hive, nectar-bearing foragers disgorge their honey sacs to one or more house bees. The house bee approaches the forager who opens her mandibles and regurgitates a tiny drop of nectar at the base of her proboscis while it is folded back under the head in the normal position. The house bee stretches out her proboscis and takes the nectar, during which time there is much mutual stroking of antennae.

If there is a heavy nectar flow, the house bee may deposit her received load into any nearby empty cell and 'go back for more', but otherwise she manipulates the drop of nectar by continually folding and unfolding her mouthparts and swallowing and regurgitating. In this way the nectar is exposed to the warm atmosphere of the hive. After some time she deposits her load into an empty cell or in one partly filled with ripening nectar where the process of water evaporation continues. It has been estimated that within an hour of reaching the hive the sugar content of the nectar has increased by 45 to 60 per cent. In normally-ventilated hives the whole process is completed in three to five days. During this time the nectar sugars (disaccharides mostly) are being broken down into simple sugars (monosaccharides) by the catalytic action of the enzyme invertase which the forager has added during her flight home and which has been added to by house bees.

During a nectar flow, evaporation of surplus water is accelerated by 'fanning', i.e. groups of bees standing in the entrance and vigorously moving their wings in a flying motion thus directing a current of air through the hive. Often two groups can be seen at opposite ends of the entrance, one fanning in and the other out. To the beekeeper one of the happiest sounds (and smells, if a smell can be said to be happy) is this pleasant hum on summer evenings after a day of successful foraging, with the delightful floral perfume being fanned out. By using small anemometers it has been shown that between 500 and 800 cubic feet of air per hour is drawn into the hive in this way.

Both high temperatures and high humidity will incite bees to fan. A good deal of energy is involved and probably four to five ounces (100-150g) of honey (carbohydrate) is expended in eliminating 16 fl oz (450ml) of water. The drying power of the incoming air is considerably increased when it becomes warmed to the temperature within the hive. In spite of what is said above, or rather as a corollary of it, a certain degree of humidity must be maintained within the brood nest to ensure that the larvae do not become dehydrated.

Temperature Control
Temperature at the centre of the brood nest is remarkably

constant. The brood can only develop between 90 and 97°F (32-36°C) and it is clear that wide outside fluctuations have little effect on the centre of the brood nest. Towards the edges of the nest the temperature is lower and less constant than at the centre. There is no evidence that bees produce heat especially to maintain the correct temperature, unlike birds whose body temperature is higher during the incubation of their eggs. In any case, bees are cold-blooded creatures. However, this is really a misnomer. What happens is that the bees' body temperature rapidly assumes that of the immediate environment, so that if a temperature higher or lower than the ambient air is required the bees have to change the temperature of their environment by adjusting the tightness or looseness of the cluster.

The arch of sealed stores normally found round and above the brood nest plays its part as an insulator. Heat is a by-product of the day to day activities of the bees and the digestion of food. It should not be forgotten that the larvae too produce a good deal of heat while digesting food.

In very high outside temperatures bees bring in water which they deposit in empty combs or on any convenient horizontal surfaces and the vigorous fanning described above results in further temperature reduction by the cooling effect of evaporation. During a nectar flow there is sufficient evaporation taking place in all but very high outside temperatures. It is only when little or no nectar is coming in that water for evaporation is needed.

The shape of the brood nest is a factor in heat retention and regulation. It has been observed that when there is ample lateral and vertical space, wild colonies seldom exceed five combs wide and more often are content with three. The centre comb often has a very long patch of brood, sometimes as much as 4 ft (120cm) by $1\frac{1}{2}$ ft (45cm). It would seem that warmth is more readily retained and regulated on a few combs with long vertical axes. Could this possibly be an argument for tall brood boxes (or a combination of boxes) with correspondingly narrow widths?

When the hive temperature falls to around 64°F (18°C) a cluster begins to form and becomes compact when the temperature falls to 57°F (14°C) and below. It has been noticed that at the onset of cluster formation many bees feed actively rather in the way they do before issuing as a swarm. Heat loss by conduction is negligible as beeswax is a very poor conductor of heat, and most loss is by convection and radiation. The loss by radiation is proportional to the surface area of the cluster. As the temperature falls even further, a tight cluster will also prevent heat losses by closing the gaps between individual bees and thereby reducing convection.

Temperature at the centre of the cluster is remarkably constant in spite of outside cold. In a normal colony this lies between 68 and 86°F (20-30°C), rising a little during long periods of confinement (it is thought that the accumulation of faeces in the rectum acts as an irritant) and falling again after a mild spell and a good cleansing flight. Temperature at the centre must rise to between 90 and 95°F (32-35°C) for brood-rearing to take place with the optimum being about 94°F(34.5°C).

Comb Building

When new combs are required, for example by a swarm, wax-producing workers hang in festoons as already described and, having produced their wax scales and kneaded them, start by applying a thick ridge of wax to the upper surface of the cavity in which they have settled. From this they extend a mid-rib downwards which will form the centre of a double row of cells. Cell walls are drawn out laterally sloping slightly upwards from the mid-rib. The starter comb is heart shaped, the vertical axis being longer than the horizontal. The edges of the cell walls are thickened to form a sort of coping and the growing tip of the comb is also made of thicker wax. The surplus wax in the thickened edges is utilized for cell-building as the work proceeds.

One is tempted to ask; why hexagonal? and why are the cells always the same size (or as nearly the same size as the human eye can distinguish)? The cells are hexagonal because, as mathematicians and engineers will confirm, this is the shape of container which will hold the greatest amount of honey, gives the greatest rigidity and involves the least expenditure of wax. Moreover, it is the shape assumed by natural forces. If a small quantity of soap or other detergent is beaten up in water, the bubbles will assume a roughly hexagonal shape where the walls of the bubbles touch. As far back as 1859, Darwin pointed out the transition from round cells to hexagon in different races of bees. Bumbles build a collection of individual round cells. The Mexican stingless bees *(Melipona domestica)* also build special cells but they are placed in a haphazard fashion. Where they touch, the dividing walls become flat and one wall will be common to contiguous cells which then assume an angular shape. The more highly-developed honey bee has carried the process to its logical conclusion.

Comb building is not a continuous process but is normally carried out only when new comb is needed. However, there is one aspect of wax manipulation which is fairly constant throughout the active part of the year. This is cell-capping. During larval development the top edges of the cells are thickened as already described so that a reserve of wax is available right at the place where it will be needed. There is also the question of reinforcement. Comb which has been in use for some time is always found to have an admixture of pollen and propolis. The proportion of propolis is always greater in the higher parts of old comb where greater strength is needed to support the weight of comb below. This strengthening with propolis is particularly noticeable in natural comb—less so in comb built on modern wired foundation.

Every beekeeper must have noticed that the wax scales issuing from the bees' wax glands (and new comb freshly built) are quite white but soon acquire the familiar golden yellow colour. But where does the colour come from? We can discount the effects of incorporated pollen and propolis because these can be removed by filtration, but the colour remains. Various beekeepers and researchers have suggested different and not entirely satisfactory answers. We know that cells are varnished by young workers, and indeed the whole of the inside of the hive will be covered with a waterproof varnish. It is not known for certain whether this 'varnish' is a glandular secretion used in the pure state or whether it is a solution of propolis dissolved in a glandular secretion. On balance the evidence points to the latter explanation. During a heavy flow from sainfoin the combs and all parts of the hive become heavily stained a deep yellow. Oil globules are present on the surface of many species of pollen grains and sainfoin pollen has particularly large and deeply-coloured globules. It is at least probable that the yellow colouring of wax is explained by the absorption of this coloured oil into the wax.

Drone Comb

Many beekeepers regard drone comb as useless and wasteful of effort and therefore attempt to get rid of all of it from their hives. I wonder if this is altogether wise. It does seem that bees like to provide some drone comb for possible future needs and it has been my observation that the presence of drones during the active season is a factor leading to good colony morale.

In natural conditions colonies always build some drone comb within the brood area—usually at the lower corners of combs. Indeed, wild colonies very often have quite large areas of such comb, so it seems that the instinct is to see that there is always drone comb available. At the same time it is true that small colonies will not build drone comb and that the rearing of drones is closely regulated by the workers throughout the summer months, drone eggs and larvae being removed as and when the colony 'considers' this necessary—and the expulsion of drones in late summer is well known. Possibly the factor which initiates this is the cessation of nectar flow. An extension of flow, due to abnormal weather conditions or artificially created by feeding will delay the expulsion of drones. Lacking a queen will do the same.

Brood-rearing

It is generally accepted that in springtime colonies with adequate food stores (at least 20 lbs/9kg plus) rear more brood than those with smaller quantities of stored food, even if the latter are supplemented by feeding sugar syrup and even if the supplementary feeding is continued so that the colony is never allowed to be short of food. Recent experiments seem to indicate that feeding with sugar syrup in spring results in very little increase in brood-rearing and may even reduce it. If a colony is desperately short of food and in danger of starvation in the near future, then food must be given of course. This seems yet another argument for the old beekeepers dictum 'do your spring feeding in the autumn'.

Nectar flows during the active season sometimes have a restrictive effect on brood-rearing. A heavy nectar flow blocks part of the brood nest with fresh nectar in the process of being processed into honey, giving the same effect as a colony given insufficient combs. This kind of brood restriction varies with the strain of bee; for instance Caucasian strains will often fill a brood chamber with nectar so that very little room is left for the queen to lay in. On the other hand, Italian strains, given super room, will increase brood-rearing on a nectar flow. There is also evidence that in most strains, egg laying increases most vigorously after a period in which it has been restricted for one reason or another. Winter temperature is an obvious example, but it is probable that heavy nectar flows—and the periods after such flows—have a similar influence since cells which otherwise would have been free for egg-laying are used as dumping grounds for incoming nectar as has been noted above.

The converse seems to be true in the case of pollen, a good intake of which has a stimulating effect on brood-rearing. The answer is to give added storage space (supers) in advance of expected nectar flows. This may sound like a counsel of perfection, but it does not take the beekeeper many seasons to know when major (and some minor) flows are to be expected. Of course his careful calculations will often be upset by the vagaries of local climate. It is impossible to over-emphasize the importance of ample supplies of pollen being available in spring. Colonies with little or no stored pollen literally stand still in spring while colonies of comparable strength with plenty of stored pollen romp ahead.

One reason for the importance of pollen in brood-rearing is that the nurse bees' pharyngeal glands cannot reach full development without plenty of protein, and pollen is the only source of this available to bees. If the pharyngeal glands remain undeveloped they cannot produce brood food. Fortunately, pollen starvation is not common in the United Kingdom, but it does occur from time to time and it is possible that cultivation of large areas of certain crops combined with the grubbing up of old hedges may make this more common. The shortage is sometimes remedied by giving the bees pollen substitutes. Various substances have been tried—dried milk, white of egg, dried yeast, soya flour, potato flour, the residue left after the distillation of alcohol from maize are some of them. All, I think, are rather poor makeshifts, but probably the most satisfactory, or the least unsatisfactory, is a sort of soft cake made of four parts of soya flour and one part of dried yeast moistened to the consistency of putty with a little sugar syrup. The cake is put immediately over the combs on the top bars.

Maturation from Fertilized Egg

It has been known for a very long time that a colony, or part colony, finding itself without a queen, will very quickly take steps to raise a new one providing it has eggs or young larvae. The mechanism which triggers this off and the advantages of it to beekeepers are discussed elsewhere in this book, but here we will consider the intriguing question of the differentiation between queen and worker. Both arise from similar fertilized eggs, but the adult insects have totally dissimilar characters physically.

Workers develop in the familiar hexagonal cells which form the normal honeycomb. They are approximately $\frac{1}{5}$ inch (5mm) in diameter and lie horizontally. Queens develop in much larger acorn-shaped cells $1-1\frac{1}{4}$ inches (25-32mm) long hanging vertically from the bottom and sides of a comb or, in the case of cells raised in emergency conditions, from occupied worker cells on the face of a comb. In this event the worker cell is enlarged outwards and downwards. The cells have a characteristic reticulated appearance externally.

In their large cells the queen larvae are given an abundance of brood food, so much that a surplus remains in the bases of the cells even after the larvae pupate. This heavy and continuous supply of food is continued right up to (or until just before) the moment of the onset of pupation.

The worker larvae, on the other hand, are given this surplus (called 'mass provisioning') for three days only. The quantity is then restricted and the larvae are given frequent meals, each small in quantity, according to its needs. The brood food is also diluted with nectar and/or honey.

The food of older larvae contains pollen. It used to be believed that the pollen was deliberately added by the nurse bees, but researchers (e.g. Haydak, Simpson) have cast considerable doubt on this and it is now thought that the inclusion of pollen is accidental via the nectar and honey. Traces of pollen have been found in the food of queen larvae.

Apart from such quantitative differences there are considerable qualitative differences between queen and worker larval food, as we have seen. The carbohydrate content of the latter exceeds that of the queen's brood food, its protein content is less and the fat content is considerably less.

It used to be thought that this abundance of brood food on the one hand and the adulteration on the other were the sole reasons for the differentiation between queen and worker. It is hard to believe that such fundamental differences between the two castes are so simply explained and indeed careful work and close consideration of the facts have led to the conclusion that this is not the case. It has been noticed that there is a significant difference in the rates of the vital processes between queen and worker larvae when they are only twelve hours old after hatching, when both have had mass provisioning of brood food and before any adulteration has taken place.

Although feeding is important and should not be overlooked, there surely is some other controlling factor. I think it is fair to say that the complete answer is not yet known (and this is the charm and attraction of many facets of beekeeping), but it is at least possible and some scientists have suggested that there are hormonal additions, via glandular secretions, to brood food which may have far reaching effects. There is a chain reaction to be considered too. The rapid development of juvenile ovaries in the queen larva, which does not happen in the worker larva, may have its effect. The juvenile ovaries may themselves release secretions which circulate in the blood of the larva and inhibit the development of worker characteristics and promote those of a queen. In recent years much work has been done by cell biologists on the organizational potential within the living cell and it must be the case that such latent powers lie within the fertilized egg, whether it be from bee or man, and this will contain the blueprint of the maturing animal.

Queen Recognition within the Colony

A laying queen receives special attention from the workers in her colony. She is surrounded by a ring of young workers, twelve to twenty in number all standing with their heads towards her. Beekeepers call this a 'court'. As she moves across the face of the comb or on to neighbouring combs the court goes with her. The court is not composed of the same individuals all the time, but as she moves about individual bees drop out and are replaced by others—invariably nurse bees. The attendants

Figure 11.   A queen and her attendants.

before the queen's head frequently feed her, probably at about half-hourly intervals. The others constantly stroke her with their antennae and lick her, obtaining minute quantities of an aromatic oily matter (queen substance) which is excreted by a pair of glands situated close to her mandibles. If a queen is removed from the colony, either by accidental loss or by some management operation of the beekeeper, the colony becomes aware of the loss, sometimes quite quickly, but in any case in a matter of a few hours. The bees become agitated and run about the combs in a disturbed manner. On meeting, bees cross antennae and the disturbance quickly spreads to the whole colony which initiates the building of queen cells.

The bees' awareness of queenlessness and their resultant agitation seem to be due to two causes, i) odour (the absence of the scent of the queen) and then later ii) the absence of queen substance. Of the two, lack of queen substance is the most important. It has happened all too frequently that a queen will unfortunately be crushed and killed during a manipulation. If the bees have access to her corpse they will continue to lick it for a little while with no sign of agitation, but this will only be temporary. Experiments where the queen is confined by a double screen so that the whole colony can smell her, but only part can touch and feed her, showed that the part which could touch and feed her continued happily, but the part which could smell but not touch her soon exhibited signs of queenlessness. If the screen was only single so that both parts could have physical contact through the mesh, neither part showed any sign of queenlessness.

Laying Workers

All worker bees have rudimentary ovaries consisting of three or four ovarioles—long, tapering, transparent tubes attached at the fine end to the heart at the forward end of the abdomen and reaching back to the sting chamber. In normal bees these ovaries are extremely difficult to see in dissections. The material of which they are made is so tenuous and transparent that they are virtually invisible in the body fluid or the dissecting medium. However, in conditions of queenlessness, when it has not been possible for the bees to raise a new queen and get her mated, some of the workers will undergo ovary development and will begin to lay eggs in a random pattern. Since these eggs are unfertilized, they develop into stunted drones, limited by the smaller size of the worker cells in which they grow. Such colonies will try unsuccessfully to raise queen cells so they appear to remain aware of their queenlessness. These colonies are difficult to requeen. Introduced queens are killed and a comb of young brood and eggs from another colony will often be ignored.

Other races of bees, e.g. *A. indica* (India) and *A. capensis* (South Africa) seem to tolerate laying workers as a normal part of colony life.

Swarming

This is the normal method by which a colony reproduces itself. It is important to bear in mind that the *colony* is the unit, not the individual bee. The subject is of such importance to the bee-keeper that we will deal with it in depth elsewhere. Suffice it to say here that it is a vital part of colony behaviour. The instinct to procreate is one of the most powerful in nature and without it the species would die.

Division of Labour

In summer a good colony of bees may consist of up to 80,000 workers, a few hundred drones and be headed by a fertile laying queen. Within the hive there will be patches of unsealed brood, both young and older to be fed and attended to, damage to comb to be repaired, possibly fresh comb to be built, incoming nectar to be received and be processed into honey, pollen to be stored, debris and dead and dying bees to be carted outside, empty cells to be cleaned, the hive to be defended against marauders, an air-conditioning system to be regulated and manned, etc. Without some system of division of labour much energy would be wasted and chaos would prevail.

Over the thousands of years of evolution bees have developed a system which meets the situation and is, at the same time, flexible and adaptable. On emergence, the newly-hatched bee cleans and grooms herself and does little other than get herself dry and her wings expanded and hardened. She wanders about the combs and frequently solicits food from other passing workers. She thrusts her extended tongue towards them until she finds one willing to feed her. The donor opens her mandibles and regurgitates a drop of food from her honey sac which the young bee eagerly sucks up. During this time there is a good deal of mutual stroking of antennae. This drop of food consists of a mixture of nectar and honey with traces of queen substance either obtained directly from the queen or from the queen via other attendant workers.

Soliciting for food like this goes on for three days while she gathers strength and rests (and thereby helps to incubate brood) and cleans vacated cells ready for more eggs. After this she is ready to participate fully in the life of the colony and helps herself to honey and pollen from the colony stores. Her age determines the type of work she is likely to undertake unless an emergency situation arises.

Table 4.   The Ages at which Various Duties are Undertaken

Age in Days	Work Likely to be Done
1-3	cell cleaning
3-6	feeding older larvae
6-15	feeding younger larvae
10-23	hive-cleaning and other house duties
10 and on	first flight from hive
12-15	receiving nectar from returning foragers
16-18	wax-production and comb-building
20 and on	foraging
24 and on	guard duties

However it must be stressed that the above dates are variable. The system is very flexible indeed and the overriding factors deciding who does what are the needs of the colony at the time. If necessary, young bees can and do undertake field duties when only a few days old and old bees will secrete wax, build comb and feed brood if the need arises.

The question which immediately arises is who or what directs the workers to their various jobs? Many fanciful suggestions have been made as to the existence of a 'hive mind' and other romantic ideas. One charming one is attributed to Aristotle (normally an accurate observer). He tells us that at dawn one particular bee makes a loud buzzing sound within the hive two or three times 'in order to arouse the other members of the colony from their slumbers and send them forth on the duties of the day'.

I would dearly like to swallow this one, but with the best will in the world cannot do so. It seems to me that the answer is more pragmatic and not quite so simple. Incoming food is rapidly shared among the occupants of the hive so they become aware of the quantity and variety of the incoming food found by the foragers. All living organisms, and bees are no exception, react to external stimuli and quickly become aware of shortages and excesses. Shortages give rise to feelings of hunger. For instance a general shortage of pollen will give rise to a pollen 'hunger', increasingly so when much protein is needed during periods of heavy brood-raising; crystallized honey in stores, a need felt for water to dissolve the honey; heat and humidity, a need felt for coolness, dryness etc. I think bees react to these needs and set about remedying them instead of, as we humans do, form committees to discuss interminably what should be done by someone else. The system is self-regulating. As one need becomes satisfied less and less workers carry out the particular task and divert to other jobs.

There is one exception to this. The presence of a nectar flow will *always* be foraged and successful returning foragers will recruit more and more workers to work the source of food until almost every cell in the hive is packed with honey. In exceptionally good nectar flows space for the queen to lay is often severely restricted. The reason for this is the success of bees in surviving in a hostile environment. Bees will *always* collect food when it is available. They have this urge to collect as much food as possible in good times and store it in a way in which it will not deteriorate until times of dearth. The colony is almost immortal. Its constituent parts are all replaceable—combs, workers, drones, even the queen—so that barring accidents, either physical or of disease, the colony goes on as a living entity for a very long time indeed.

Figure 12.   The male or drone honeybee (*Apis mellifera*).

Figure 13.   The worker honeybee (*Apis mellifera*).

Figure 14.   The queen honeybee (*Apis mellifera*).

Foraging

From spring to late summer scout bees issue from the hive and search around for possible sources of nectar, pollen, propolis and water. For the sake of simplicity we will take the search for nectar as an example—nectar is the most important factor in the search for food. The scout bee is attracted to nectar sources by both sight and scent. Of the two scent is the most compelling influence. The scent of lime trees is detectable by the human nose at least 30 yards away. How much greater will be the distance in the case of the bee with her more highly developed olfactory sense. Having been attracted to a possible source by scent, the final approach of the scout is guided by the shape and colour of the flowers. The scent gradient, steeper as the scout homes in on the patch of flowers, will increase the excitement and the promise of good things to come.

Individual flowers in many species have what are called 'scent guides, on their petals. These are lines of scent on the petals increasing in strength as they lead down to the nectaries. The odour of the scent guides is often different from the general odour of the flower. All these factors help to guide searching bees towards nectar sources and down to the actual nectaries. In doing so the bee is guided past the anthers, from which she may or may not want pollen, but in either event she cannot help but gather pollen on her hairy body—to the mutual benefit of both plant and bee.

Having found a good source of nectar the scout fills her honey sac and returns to the hive. On arrival, she alerts other bees and recruits them to forage the source she has discovered. She does this by means of the bee 'dances' so well described by Karl von Frisch. She cannot convey the visual appearance of the floral source she has found but she can, and does, pass on information as to its scent, direction and distance. She dances on the vertical face of a comb, or more rarely, on the alighting board, and her excitement attracts the attention of other workers who follow her movements.

They can perceive the scent which clings to her body and this is a guide for recognizing the source outside the hive. Distance is shown by the type of dance. If the source is not more than 50 to 100 metres, scent is enough indication and the dance consists of a tight circle with a diameter of about two bees body lengths. The direction of the circle is frequently reversed. For greater distances the dance becomes a sort of flattened figure of eight—two more or less circular movements with a straight run between. As the dancing bee travels up or down the straight part she waggles her abdomen from side to side. The tempo of the dance indicates the distance of the source from the hive.

Von Frisch estimated that 9 to 10 complete dance cycles every 15 seconds indicate a distance of 100 metres, 8 to 9 cycles 200 metres, 6 cycles 500 metres, 4 or 5 cycles 100 metres. In other words the farther the source, the slower the dance. While dancing, the successful scout emits a buzzing sound which may convey information too—or it may merely be to call attention to herself. The direction of the source too is given by the straight run of the dance. The angle this makes with the vertical is the same as the angle between a line to the source and a line to the

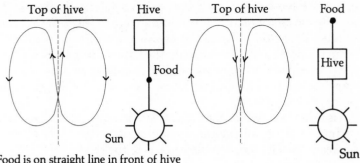

Food is on straight line in front of hive

Food is on straight line in back of hive

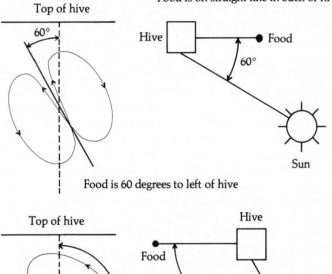

Food is 60 degrees to left of hive

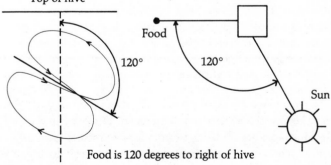

Food is 120 degrees to right of hive

Figure 15. Bee dances indicating sources of food. When performed on the face of a comb the angle of the dance to the perpendicular equals the angle between the sun and the source of the food.

sun. We have seen that bees can perceive polarized light and it is an extraordinary fact that providing there is only a small patch of blue sky visible, bees can be aware of the direction of the sun from the polarized light reaching them from the blue sky. There is an intermediate dance for middle distances which is a round dance opened out a little to a crescent shape rather like the wings of a sycamore seed.

As other bees get the message, they too go out, find the source of food and on return perform their dance and recruit more foragers until a considerable force is in the field.

Pheromones

In recent years attention has been focused on the effects produced by infinitely small quantities of substances produced in the glandular systems of different organisms. These fall into two main groups. First there are hormones which circulate within the body of the individual producing them and have a regulating effect. If the hormonal balance is disturbed or the supply inadequate the body will malfunction and may die.

The second group comprises the pheromones. These are similarly produced by and in an organism's glands, but are emitted externally. They convey information and elicit responses and reactions by other individuals of the same species.

We have noted the powerful effect on a colony of bees of the supply of queen substance and the equally remarkable effect of the lack of it. Queen substance is a pheromone, and there are others. When a bee stings, an alarm pheromone is released which puts other members of the colony on the alert and can result in mass aggression. This is why you will sometimes see an experienced beekeeper who has collected a sting blow smoke over it. He is attempting to mask the odour of the alarm pheromone.

A pheromone is used for trail marking by a successful forager to guide her sisters to a source of food. The *Nasonov* gland at the tip of a workers abdomen emits a pheromone used as a calling signal when bees are disturbed or in a fresh location, e.g. a newly-found site for a swarm. It is recognized by other members of the colony and homed in on. It is possible that a pheromone is released by young brood when they are fed which reacts on the nurse bees.

Much work remains to be done on the chemical stimuli which control and direct insect behaviour and it is at least conceivable that future research may discover methods of pest control other than by means of lethal poisons.

6

Pollination

Pollination is the term used to describe the transfer of pollen from the anthers of flowers (the male element) to the stigma (the female element) of the same flower, or other flowers of the same species. This is not fertilization. That occurs when pollen tubes grow down the stigma from the implanted pollen grains to the ovary and the cell nuclei from pollen grain and ovary fuse to initiate ovulation.

Many plants are air-pollinated and have no need of insect agents. Coniferous trees and all the grasses (and this includes all cereals) fall into this category. The pollen is very light, is released in enormous quantities and is carried on the slightest breeze. The pollen grains are usually smooth and some, pine for instance, have enlarged sacs rather like balloons to facilitate air travel. In contrast most pollen grains from plants needing insect pollinators have their surfaces covered with spines, are deeply etched or carry filaments—all devices to enable the grains to cling to the hairy bodies of their insect visitors. In beekeeping we are, of course, concerned with the last category for it is these plants which offer other inducements attractive to bees—nectar for instance.

'Self-fertile' plants are those which are able to produce viable seed with their own pollen, but there are many so-called 'self-sterile' plants which cannot produce viable seed or fruit set unless pollinated from a different plant and sometimes from a different variety of the same species. Examples of fruit trees which require special pollinators will readily spring to mind. Even in plants which are self-fertile, cross-pollination seems always to give improved results, probably because of the introduction of hybrid vigour.

Many insects visit flowers for pollen and/or nectar, but few are consistently good pollinators and none other than the honey bee, can be concentrated in large numbers at a particular crop at the right time. This is for the obvious reason that the modern system of beekeeping makes it possible to move complete hives at night after flying has ceased and place them in orchards and among field crops just at blossom time when the bees are needed and to remove them when they have completed their job of pollination. I believe there is a case for arguing that the value of bees to the agricultural industry at least equals the value of the honey they produce and may well exceed it. A recent report of the National Agricultural Advisory Service observed that 'It no longer pays to grow winter or tick beans in the East Midlands without the use of bees. The average increase in yield over six years of trials showed that bees increased the yield by eleven cwts per acre.'

The situation in recent years has been aggravated by mechanized farming. Headlands are ploughed up and old established hedges are grubbed out to make it easier for tractors to work single crop fields of large acreage. These methods have destroyed the natural habitat of wild insects, especially wild bees, which used to play an important part in pollination. The need for honeybees has therefore increased.

The farming community has understood this and a useful co-operation has grown up between grower and beekeeper. Many beekeepers devote some of their management methods to the provision of colonies specially prepared for pollination and these are hired out to growers. Either written or implied, an arrangement is come to under which the beekeeper agrees to move colonies of bees to where they are needed. He undertakes to see that the colonies are vigorous and healthy and the grower undertakes to notify the beekeeper of the commencement of flowering and also the cessation so that the bees can be moved out. He will also notify the beekeeper of his intention to spray the crop with insecticide so that avoiding action can be taken. It is important that the beekeeper keeps to his side of the bargain, that is, that he supplies strong, healthy colonies at the time he is asked to. A small colony covering about five frames in the hope that it will build up on the crop will not do. One or two points are worth noting. Colonies should be distributed throughout the orchard or field if possible and not grouped together.

Better pollination will result if bees are not taken to the crop until it is just in flower. If taken before blossom opens there is a risk that the bees will find other sources to forage and may remain faithful to them. Large colonies with plenty of brood will have a greater need for pollen than small colonies with little brood. Colonies fed with syrup in inside feeders will often divert nectar foragers to pollen. The provision of clean water close to the hives will save the bees time and effort.

The list overleaf gives some indication of the dependence on insect pollination of some of the more commonly grown crops for viable seed and fruit set. With the exception of peas and some rapes all the crops benefit by cross pollination, even when it is not essential, as the example of field beans shows. Effective insect pollination, as well as increasing both the quantity and quality of the crop, tends to make the plants mature earlier. They are shorter in the joints making mechanical harvesting easier and lessening the danger of storm damage.

Table 5.   Insect Pollination Requirements of Plants Grown for the
Market or for Seed Production

Crop	Need for Pollination
Apple	E
Field bean	S
Runner bean	E
Blackcurrant	E
Blackberry	E
Cabbage	M
Cherry (sweet)	E
Clover (red)	E
Clover (white)	E
Gooseberry	S
Onion	E
Pear	E
Plum	E
Garden pea	N
Rape *(Brassica napus)*	N
Turnip rape	E
(Brassica campestris oleifera)	
Raspberry	S
Strawberry	S
Tomato	E

Key:
E = Essential   N = not needed
M = Moderate need   S = small need

7

The Anatomy of the Bee

Books dealing in detail with honeybee anatomy are listed towards the back of the book (see *Further Reading*), but a brief glance at the main anatomical features of our bees will be useful here.

The Exoskeleton

Vertebrates have developed a bony skeleton which is light compared with the total body weight, is strong, gives good anchorage points for the muscles and gives some protection to the soft viscera. By contrast bees (and other insects have evolved an external skeleton consisting of a number of layers. The outside layer is waterproof to prevent dehydration and is covered with plumose (branched) hairs. It bears a large number of sense organs in addition to the eyes consisting of pegs, depressions and plates.

The main constituent of the cuticle, as the outer shell is called, is chitin, one of the most durable substances in nature, perhaps equalled only by pollenin, the outer surface of pollen grains. There are thickened ridges internally (phragmae) and internal peg-like extensions (apodemes) which serve as anchorage points for the muscles and internal struts which give rigidity where wanted, e.g. in the head.

The whole exoskeleton consists of plates or segments, some fused together as in the thorax and some overlapping and connected by thin flexible membranes which give flexibility. This is most clearly seen in a bee which comes in after a flight. The abdomen contracts and expands longitudinally. She appears to be panting and is in fact driving air in and out of the spiracles down the sides of her abdomen. After she has rested awhile these movements cease, so perhaps 'panting' is a fair description. Parts of the exterior are so specialized and altered that it is not always apparent that the bee is a segmented animal —an attribute inherited from simpler worm-like ancestors. There are in fact nineteen segments distinguishable; six in the head, four in the thorax and nine in the abdomen.

The Head

The head is a flattened box roughly triangular in shape. The segments are fused together to make the box rigid. A pair of mandibles are hinged above the mouth. They are strong, concave and ridged on the inner faces. They are used for biting, cutting, kneading wax, handling propolis, transferring pollen into the mouth (but not crushing pollen as is sometimes supposed), feeding larvae, dragging debris from the hive, grooming, fighting and are the bee's all-purpose tools.

The proboscis is slung on hinges below the mouth. Although it is extendable to about 6.5mm in the case of the dark European bee and up to 7.2mm in the Caucasian bee, it is hinged along its length so that it can be folded out of the way below the head when not in use. It is a complicated structure consisting of a number of parts which can be held together to make an air-tight sucking tube.

Eyes

Three simple eyes (ocelli) are sited towards the top of the head. They are not capable of forming any image on a retina and it is thought that their function is to determine the intensity of light —and also possibly to see which way is 'up'.

The compound eyes, on the other hand, are extremely

complicated. Each eye consists of some 6,900 facets in the worker and 8,600 in the drone, each about 0.2mm in diameter. Behind each facet is a lens followed by a crystalline cone reaching down to the retina. Each cone is separated from its neighbours by opaque pigment. From this construction the retina cannot form an image as in the mammalian eye, but receives a mosaic of dots of varying intensities. Visual acuity is poor by our standards, but detection of movement is excellent. Bees can detect flicker up to about 54 per second: in man the maximum is about 30—as the film industry knows.

The Antennae

Two antennae are located on the face between the compound eyes. Each consists of a rigid rod, the *scape*, attached to the head by a ball and socket joint giving all round movement, followed by a jointed rod, the *flagellum*, about twice as long as the scape. The flagellum is covered with many thousands of sense organs of different kinds—pegs, hairs, and plates. It has been estimated that there are probably as many as 500,000 sense organs in the single antenna of a drone. The plates are particularly numerous towards the tip of the antennae. It has been said that they are organs of taste and smell, but it is far from clear how they function. They are certainly detectors of some kind. When two bees are seen to meet and to go through their routine of antenna crossing and stroking it is hard not to believe that some kind of communication is taking place although we may not be quite sure what form this takes.

Just where the flagellum joins the scape a ring of sensory cells surround the join. This is believed to be sensitive to vibrations in the antennae and to be a speed of flight indicator analagous to the pitot tube on an aircraft.

Brain and Nervous System

To say that the bee's brain is smaller and less complex than our own would be a glimpse of the obvious, but in proportion to its size the bee has infinitely less brain capacity for reasoning or learning. There is no evidence to suggest any conciousness or intelligence, although bees have a limited but short-lived memory. There is a nicely balanced system of co-ordination to deal with responses to stimuli received via the sense organs from outside or from inside via hormones. Beekeepers are good or bad according to their ability to provide the right conditions and stimuli to enable them to carry out their systems of management.

Most of the brain consists of two large optic lobes and two smaller, but significantly large, antennal lobes which receive messages from the compound eyes and the sense organs on the antennae respectively. The nervous system consists of seven ganglia, two in the thorax and five in the abdomen, plus one large ganglion in the head, the sub-oesophageal ganglion, which is so closely attached to the brain that some writers regard it as part of the brain. Twin nerve trunks connect the ganglia and from them networks of nerve fibres run to all parts of the body (or should it be *from* all parts—it is a two way system). The ganglia correspond roughly to the segments although some are fusions of smaller segmental ganglia. There is a good deal of decentralization and the segmental ganglia control movements of their parts, e.g. legs to a great extent but it seems that the brain still has a co-ordinating influence on the whole. A bee with its head cut off will move its legs quite vigourously, but the movements are random, being deprived of the co-ordinating influence of the brain.

Wings

All the organs of locomotion are attached to the thorax. Some of the segmental plates of the thorax are fused to give a rigid box and there is a membraneous 'gusset', the *scutal fissure*, between two rigid parts of the box which allows the whole thorax to be distorted slightly to allow for the operation of the wings.

The fore wings only are driven by the flight muscles. The smaller hind wings merely trail behind and are connected by a row of tiny hooks, the *hamuli,* on the leading edges of the hind wings engaging with folds on the trailing edges of the fore wings. The wings are attached to both the *tergites* (dorsal or upward-facing plates) and the *sternites* (ventral or downward-facing plates) of the thorax where they overlap slightly at the scutal fissure forming a fulcrum and a point where pressure can be applied. There are two large bundles of muscle fibres running fore and aft and two more running top to bottom. These are firmly attached to the insides of the segmental plates and are so large as to occupy the whole of the thorax.

In flight the vertical muscles contract and pull the roof of the thorax down and when they relax the longitudinal muscles, which have been stretched, contract in turn causing the roof to arch a little. The timing is automatic because the effect of

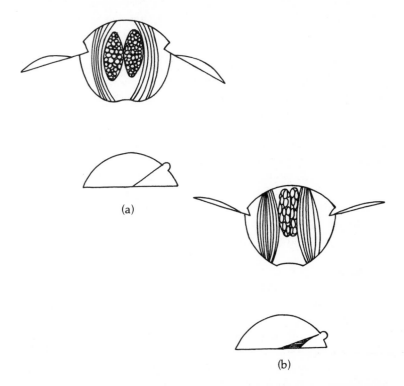

(a)

(b)

Figure 16. Operation of indirect wing muscles in flight. The drawings above show sections through the thorax. Those below represent the top half of the thorax, showing (a) closing of suture and (b) opening of suture.

(a) Longitudinal muscles contracted; roof of thorax raised; suture closed; wings depressed.

(b) Longitudinal muscles relaxed; vertical muscles contracted; roof of thorax pulled down; suture open; wings raised.

stretching a muscle is to cause it to contract immediately. Once flight has started the wing beats continue. This alternate flattening and arching of the thorax presses on and lifts the wing roots and moves the whole wing strongly up and down. The wing tips describe a shallow figure of eight and it is necessary to control the degree of slant of the wings and the 'angle of attack' in order to provide thrust—i.e. forward movement—as well as lift. Tiny movements in the attitude of the wings are controlled by small muscles at the wing roots so that the flying bee can correct pitching and rolling and other unwanted movements caused by air currents.

In flight the wings vibrate at 200 to 250 cycles per second. This is too fast to be controlled by the nervous system and this is why the power is produced by the automatic stretching of one set of muscles by the contracting of the other set.

Flight needs considerable energy and this is supplied from the blood sugar. Normally a bee has fuel for only 15 minutes flight—4 to 5 miles but can replenish by consuming nectar on the way. Some nectar collected from a source at a distance from the hive must be used for the return journey so there is a point beyond which it is not economical to collect nectar. The excess would be consumed to get back to the hive. The Law of Diminishing Returns applies to bees too!

Legs

The three pairs of legs are attached to the thorax and are used to perform a number of important functions in addition to that of locomotion.

In walking the bee always keeps three feet on the surface and moves by swinging forward two legs on one side and one on the other alternately. She always has a firm tripod supporting her and balance, as in man, is not needed. Although the legs are equipped to perform different functions, all are of a similar segmented construction. Starting at the point of insertion in the thorax the parts are: coxa, trochanter, femur, tibia, tarsus (subdivided into a basitarsus and four smaller tarsomeres) and a pretarsus bearing the feet. The joints between the segments are hinges operating in one plane only. They are in no way comparable to the ball and socket joints in common mammals.

The limitation of movement occasioned by the hinged joints is compensated, in part at least, by the joints working in different planes.

The front legs are smaller than the other two and are set close behind the head. They bear the antenna cleaners. At the proximal end of the basitarsus (proximal—towards the point of attachment: distal—the part away from the point of attachment) is a semicircular notch lined with stiff bristles. A peg-like projection on the distal end of the tibia can be closed across the mouth of the notch. In use the flagellum of the antenna is placed in the notch and the peg closed across it. By drawing the antenna through the notch between its bristles and the scraping edge of the peg, pollen grains and any other debris are brushed from it.

The feet of all the legs are similar. They bear pairs of strong curved claws and a number of stiff spines which give grip when walking on rough surfaces. Between the claws is an *arolium*, a piece of soft tissue normally kept folded on itself in the form of a shallow 'U'. When on a smooth surface, offering no foothold to the claws, these are spread wide apart and the 'U' unfolded so that the arolium is pressed against the smooth surface to which it adheres. Quite how it does this is not clear. It has been said that a sticky fluid is exuded from the stiff spines on the feet but close dissection and examination have failed to detect any such fluid or any organ from which it could originate.

The middle leg has no special tools. A single spine is borne on the distal end of the tibia. It used to be said that this was used to detach wax scales from the wax mirrors, but this has never been observed and it is thought that the spine has no function.

The inner sides of the front and middle legs are partially covered with stiff hairs which are used as brushes to clean the thorax and head of pollen and to pass the collected pollen back to the hind legs.

The hind legs bear the apparatus for collecting pollen and transporting it back to the hive. The basitarsi are broad and flattened. On the inside faces there are rows of downward-pointing stiff hairs. These are used to brush pollen from the abdomen during flight. The pollen collected by the front and middle legs is moistened with a little nectar regurgitated from the honey sac and passed back and mixed with the pollen brushed from the abdomen. The distal end of the tibia bears a row of sharp pointed teeth, the *rastellum* or *rake*. While still in flight the hind legs are rubbed together in an up-and-down movement. The rastellum on each leg scrapes the pollen from the inside of the other leg. This is then pushed on to a sloping shelf at the proximal end of the basitarsus, the *auricle*. As the joint between the tibia and the basitarsus is closed, the pollen mass is squeezed up on to the outer face of the tibia into the *corbicula* or pollen

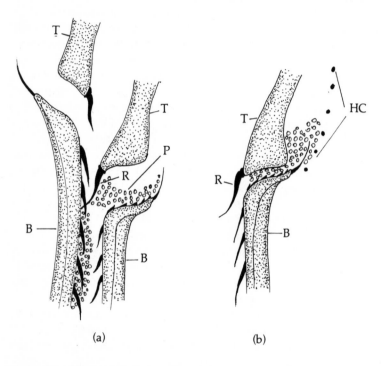

Figure 17.   Packing pollen on hind legs.

(a) Legs rubbed together. Pollen press (P) open; rastellum or rake (R) of right leg scrapes pollen from inside of the basitarsus (B) of left leg onto pollen press (P).

(b) Tibia (T) of right leg closed with basitarsus (B), closing pollen press and forcing pollen up into corbicula on tibia where it is retained by the hairs of the corbicula (HC).

basket. The corbicula is fringed with long, slightly in-curved hairs and has a single long central hair. Because it has been moistened with a little nectar the pollen mass rests safely in the corbicula on the flight home giving the familiar 'plus fours' look to the successful pollen forager. The loads are easily detached into waiting cells on return to the hive.

The Abdomen

The abdomen consists of nine segments. Only six are visible, the remaining three forming part of the sting apparatus and carrying the anus. Each segment consists of an upper plate (tergite) and a lower (sternite). The plates overlap slightly and are connected by thin intersegmental membranes giving flexibility. The front halves of four of the sternites have thinned areas—the wax mirrors. These underly their preceding sternites and have behind them wax glands. At certain times the wax glands are operative and manufacture wax which is exuded in tiny plates onto the outer surfaces of the wax mirrors.

Under the front part of the last visible tergite lies a scent gland, the so-called Nasonov gland. When calling or wishing to lay scent trails the tip of the abdomen is depressed exposing the gland which appears as a whitish strip.

The Sting

The mechanism of the sting is a little more complicated than would appear at first sight. It is composed of the very much modified last three abdominal segments. Three sets of plates are articulated together in such a way that muscular contractions drive the lancets of the sting into the victim and at the same time pump venom in. Each of two oblong plates has a flexible curved extension, the ramus, which in turn becomes a straight lancet barbed at its finely sharp end. A small triangular plate and a larger quadrate plate complete a train of levers. Protractor and retractor muscles thrust the lancets forward and withdraw them alternately. The barbs of the lancets hold in soft mammalian flesh and each set of barbs gives purchase for the other lancet to penetrate more deeply. The central portion of the sting is expanded into a stylet and a bulb. The stylet has rails which run in corresponding grooves down the length of the lancets and the

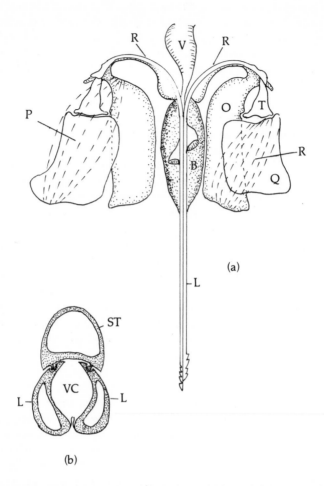

Figure 18.   Diagrams of sting apparatus. The mechanism is drawn flattened out for the sake of clarity.

(a) The protractor muscle (P), when contracted, pulls the quadrate plate (Q) towards the oblong plate (O). This causes the triangular plate (T) to swing on its articulation with the oblong plate (O) and drive the ramus of the lancet (R) forward. This in turn, drives the lancet (L), and so on, alternately left and right. V—venom gland; B—bulb of lancets.

(b) Cross sections of lancets. L—lancet; ST—style; VC—venom canal.

three parts form a slender hollow rod. The upper part of the stylet swells out into a bulb which is fed from the venom gland. Cup-shaped extrusions on the shafts of the lancets operate as pistons/valves, pumping the venom down the hollow centre and out at the tips of the lancets.

The venom gland, sometimes called the acid gland is a long narrow tube opening into a large sac in which accumulated venom is stored. The venom itself is a mixture of enzymes, mainly phospholipase A and hyaluromides. These release histamine in the body of the victim and give rise to the characteristic swelling. The venom also contains a protein poison, apitoxin, the effects of which are similar to those of cobra venom—but in very minute doses of course.

A second gland, the alkaline gland, is closely attached to the sting apparatus, but its products are not discharged into the venom bulb. Its function is not clear but it may be merely one of lubrication. The rhythmic muscular contractions when the sting is in action are controlled by the last ganglion of the main nervous chain. When the sting is driven in, the bee struggles to free itself but the barbs of the lancets hold firmly in soft flesh. The membraneous attachments to the sting chamber are weak and in the struggle they rupture and the bee flies free to live for not more than a few hours, leaving behind her the whole of the sting apparatus. The last ganglion and the operating muscles are left behind too and continue to operate, steadily driving the lancets in and pumping venom. For these reasons it is best to remove the sting from the wound as soon as possible by scraping with a knife blade, the sharp edge of a hive tool or something similar. Do not seize the sting between the thumb and forefinger and pull it out. You will certainly succeed in doing this but you will also squeeze the poison sac and the bulb and squirt the whole of the venom into the wound.

The Digestive System

Like any other animals bees need two types of food, for body building (proteins) and energy foods (carbohydrates). Pollen is rich in protein and is the only available source to the bee. The enzymes she produces are capable of digesting a number of proteins and this fact makes it possible for her to digest at least some of the protein content of the pollen substitutes beekeepers sometimes give to their bees in times of shortage. The bees' carbohydrate needs are supplied from the honey they process from nectar so that food requirements are totally satisfied from floral sources.

The composition of nectar and honey is dealt with in detail in another section. Let us here briefly note that bees add minute traces of enzymes to the nectar they collect. The enzymes act as catalysts and break down sugars with large molecules (complex sugars or polysaccharides) into the simpler sugars glucose and fructose. This process is known as 'inversion'. The simple sugars can be absorbed into the blood without further processing. The turning of nectar into honey is, therefore, a form of predigestion.

The process starts with the actual collection of nectar. Glandular secretions containing the enzymes, mainly invertase, pass down the proboscis and mix with the nectar as it is being sucked up. Further additions are probably made in the mouth. Pollen cannot be predigested in this way and has to be dealt with in the ventriculus. Although pollen is rich in fat, another potential energy source, bees are unable to digest it and the oils taken in with pollen are passed out in the faeces unaltered.

The alimentary system of an animal consists of a tube running from mouth to anus with different sections adapted to perform various functions. In the case of the bee the mouth has no teeth and no grinding or masticating mechanism but is merely a pump to suck up liquids and a space into which pollen may be shovelled by the mandibles.

From the back of the mouth the oesophagus passes as a narrow tube straight through the thorax. Immediately inside the abdomen the oesophagus expands into a collapsible sac, the crop or 'honey stomach'—a bad name because it is not a stomach at all, but simply a collecting receptacle with a valve at the posterior end, the proventricular valve. This valve projects into the crop. Four triangular lips allow the valve to be kept closed, for instance when a load of nectar is brought back by a forager, or opened to allow the passage of nectar, honey or pollen into the stomach proper, the *ventriculus*.

The four lips of the valve are fringed with fine hairs pointing backwards and an ingenious filtering process is possible. The nectar collected by foragers always contains some pollen grains and often a great many. The lips of the valve make gulping movements, sweeping pollen grains back into the ventriculus and gradually greatly reducing the amount of pollen in the crop. Not all the pollen can be removed by this filtering and some is carried over with the nectar for processing into honey, possibly adding food value to it and certainly acting as an indicator of the floral source of the honey as we shall see later.

The maximum capacity of the crop is about 100mg but the average load brought home by a forager is not likely to be more than half that amount and is often far less. About 75 to 80 per cent of the water in nectar is removed in the processing into honey and I leave it to the mathematically minded to calculate how many thousands of journeys are represented in one pound (450g) of honey (or perhaps we should now say a half kilo of honey).

Pollen and/or honey passes through the proventricular valve into the ventriculus. This is a wide tube coiled into a loop with restrictions along its length and also encircling muscles. The muscles contract in sequence causing a wave of constriction to pass down the ventriculus making a wave of peristalsis which urges the food onwards. The walls are lined with cells which are constantly breaking off and mixing with the food. The cells contain the enzyme which will digest the pollen grains through their germ pores. Examination of the pollen husks in faeces shows that most of them are intact but all are empty of contents.

At the posterior end of the ventriculus about one hundred fine thread-like tubes open into the ventriculus at its junction with the small intestine. The walls of these consist of a single layer of cells. They are the malpighian tubules and are excretory organs absorbing nitrogenous matter from the blood and passing it down to the small intestine. They are the equivalent of our kidneys. Another valve, the pyloric valve, lies at this end of the ventriculus. This regulates the passage of material into the small intestine. The small intestine is a long narrow tube pleated into six longitudinal folds so that increased surface is offered for the absorption of food into the blood.

The small intestine opens into the rectum, a flask-shaped sac capable of considerable distension so that the faeces can be retained for comparatively long periods in the case of bad weather.

The rectum bears on its surface six rectal pads. The true function of these has not been confirmed and they may be vestigial. It is unlikely that their purpose is to re-absorb water from the faeces as used to be supposed, but it is possible that they are concerned with regulating the level of salt in the blood.

The rectum ends in the anus, an opening in the tenth abdominal segment located just inside the sting chamber.

Blood and Circulation

Bees have no closed system of arteries and veins as we have. Blood fills the body cavity and its main function is to transfer the products of digestion from the small intestine to the parts which need them and to carry away waste products in solution for removal by the malpigian tubules. The blood consists of plasma or haemolymph and contains a few phagocytes which attack and ingest bacteria. There are no red corpuscles.

The heart is a narrow tubular organ lying just below the roof of the abdomen. It is closed at the posterior end and bears five sets of valves (the ostia) along its length. These are slit-like openings in the sides of the heart with lips protruding inwards. As the heart pulsates in a wave motion from back to front, increased pressure within the heart automatically causes the ostia to close near the point of constriction driving the blood forward into an aorta, a fine tube which traverses the thorax and finishes, open ended, just behind the brain. A current of blood is thus created moving forward along the dorsal and backwards along the ventral surfaces. Diaphragms of thin tissue are stretched over the whole of these surfaces in the abdomen attached only at intervals. These diaphragms also pulsate in a wave motion and do much to promote the circulation of the blood.

The Respiratory System

Almost every animal needs a constant supply of oxygen in order to live, and the more active it is, the more oxygen it needs and the more carbon dioxide it gives off. Bees are no exception to this. They are very active apart from the winter clustering period and have evolved a well-developed system of respiration which is perfectly adequate for their small bodies.

Air is drawn in through spiracles—openings in the body wall down the sides of the thorax and abdomen. These open into large air sacs in the abdomen and smaller sacs in the thorax and head. The large abdominal sacs act as bellows when the abdomen is alternately dilated and compressed. There are ten pairs of spiracles in all and all, except the first pair in the thorax, are fitted with flaps or lids which can be closed. Large air passages, the tracheae lead from the air sacs and these divide and sub-divide into smaller and yet smaller tubes, the tracheoles which finally reach every part of the body needing oxygen. The tracheae are fitted with chitinous spiral thickenings to the walls which give rigidity and prevent their collapsing.

Oxygen is supplied to the muscles and tissues by direct gaseous exchange through the fine open ends of the tracheoles. The massive flight muscles contain cytochrome—a substance which accelerates the gaseous exchange at a point where there is need for large quantities of oxygen during flight. Scientists have doubted whether this direct gaseous exchange can be sufficient to supply the oxygen needs of a bee in flight and have found that the blood contains haemocyanin, a copper compound, in solution. This may well act as an oxygen carrier and it is thought likely that the blood must pay some intermediate part in respiration although in no way comparable to what happens in vertebrates.

The Reproductive Organs

The ovaries of a mated queen are extremely large, occupying practically the whole of the abdominal cavity. Each ovary consists of about 150 tubules, the ovarioles, which start as fine threads in the anterior part of the abdomen just below the heart. They increase in size towards the posterior end. Cells constantly bud off at the fine ends and pass down the ovarioles. The cells are of three kinds: (i) egg cells proper, each surrounded by (ii) a mass of nurse cells and (iii) follicle cells. On their passage down the ovariole, the nurse cells absorb nutriment from the blood through the thin ovariole walls, gradually increasing in size.

They pass the nutriment on to the egg cell via a small opening in the anterior end of the egg cell and finally are totally absorbed in the egg.

The ovarioles open into two lateral oviducts which combine into a single median oviduct and this, in turn, becomes the vagina. Matured eggs pass through the oviducts and vagina into a membraneous pouch at the anterior end of the sting chamber. The mature egg is pearly white, about $\frac{1}{16}$ inch (1.5mm) long and about one seventh of this in diameter. The opening already referred to remains at one end. This is the micropyle through which sperms can enter for fertilization. The other end of the egg is covered with a sticky substance which enables it to stand on end in the cell when laid.

A spherical receptacle about 1mm in diameter, the

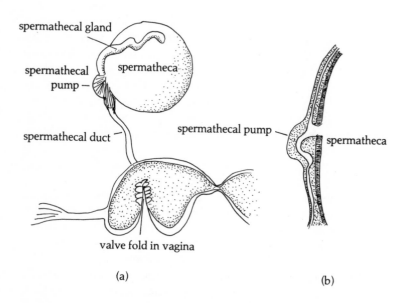

spermathecal gland

spermathecal
pump —

spermatheca

spermathecal duct

spermathecal pump

spermatheca

valve fold in vagina

(a)

(b)

Figure 19.   Details of spermatheca and vagina (a), and section through spermathecal pump and part of spermatheca (b).

spermatheca, lies above the vagina to which it is connected by a slender duct. The spermatheca is the vessel in which the spermatozoa received at mating are stored in a viable condition. Just below where the duct joins the spermatheca, there is an S-bend fitted with three sets of muscles. By contracting and relaxing in sequence these muscles create suction and then compression in the duct withdrawing a few sperms from the spermatheca and forcing them down the duct. On the floor of the vagina just below the opening of the duct there is a muscular fold, the valve fold. The queen controls the sex of the egg she is about to lay by pressing the egg against the duct with the valve fold if she wishes it to be fertilized, in which case the egg receives a packet of sperms. If she does not want it to be fertilized she allows the egg to pass down the vagina without further treatment. Fertilized eggs develop into females (workers and queens) and unfertilized eggs develop into males (drones). As is common in insects, unfertilized eggs will hatch, but as they have only a half set of chromosomes they will only develop into males. The diagram overleaf (Figure 20) will demonstrate this.

At the height of the breeding season a good queen can lay up to 2,500 eggs in twenty-four hours—some even more than this. She will lay between a million and a million and a half eggs during her lifetime. One spermatheca will hold about seven million sperms.

It is said that the queen measures the size of the cell with her forelegs and lays the appropriate egg according to the size of the cell, small (worker) cells receiving a fertilized egg and large (drone) cells receiving an unfertilized egg. If measurement is the sole factor one wonders what is the explanation of the fact that she will lay fertilized eggs in even larger queen cups. Perhaps she thinks, if small, then fertilize; if large, then leave unfertilized; if very large, then, again, fertilize?

The Drone

In immature drones, i.e. those just emerged, the testes are large white bean-shaped organs containing bundles of tubes in which the spermatozoa originate. As the drone matures the testes gradually wither away, their substance being consumed in the production of the sperms. These pass down short coiled tubes, the *vasa deferentia*, into the seminal vesicles where they remain until needed at mating. The walls of the vesicles are muscular and secrete a fluid in which the sperms lie quiescent in swathes, heads to walls. The process of maturing takes about thirteen

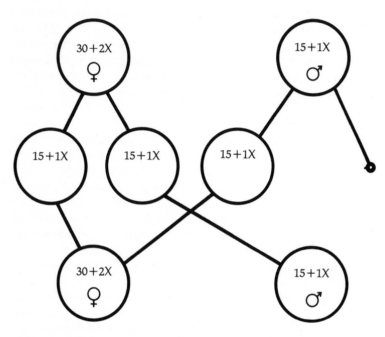

Figure 20. Egg production and normal egg fertilization giving rise to females and non-fertilization giving rise to males.

Top line. Left, queen (♀) with full set of chromosomes, 30+2X; right, drone (♂) with half set of chromosomes.

Middle line. Left, queen's eggs each with half set of chromosomes made by reduction division. Right, sperms from drone with half set of chromosomes.

Bottom line. Left, egg fertilized by sperm from drone, acquiring a full set of chromosomes and giving rise to a female insect. Right, unfertilized egg having half set of chromosomes and, therefore, capable of giving rise to male insect by parthenogensis.

days, after which time the testes have shrivelled to small greyish triangular remnants. On the other hand the seminal vesicles have grown into swollen sausage shaped organs full of sperms. The vesicles open into the ejaculatory duct where they are joined by two large mucus glands. The duct leads to a bulb and a number of complicated extrusions, the purposes of which are not known. The whole of the ejaculatory system is known as the *endophallus*. At the moment of copulation the drone's powerful abdominal muscles contract violently and this causes the whole of the endophallus to turn inside out. At the same time the muscular walls of the seminal vesicles contract and force semen down the now everted endophallus. This is followed immediately by contraction of the mucus glands and an emission of mucus closely follows the semen.

Mating

Many attempts at controlled mating in confined spaces with selected drones have been made, but all have been unsuccessful. Artificial insemination is often carried out in the process of line breeding and selected areas are sometimes flooded with drones of a known strain and used as mating areas. For instance German beekeepers for some years have maintained certain of the Frisian Islands as such drone mating stations and beekeepers on the mainland will send their virgin queens there for mating. Quite detailed pedigrees of queens can be, and are, made as a result.

Mating always takes place on the wing and never below about 35 feet (10m) from the ground. Successful matings have occurred with tethered queens above this height. Drones get together in congregation areas—often from many hives—and virgin queens seek these out on their mating flights. When they are noticed by the drones a chase ensues and the swiftest drone mates. In the actual coupling the drone mounts the queen's back and his partially everted endophallus is inserted into the queen's sting chamber. The drone falls back paralysed, eversion is completed and he falls to the ground. Semen is injected first followed by the mucus which hardens on contact with air to make a plug preventing loss of semen. The plug of mucus also holds the endophallus in place and is visible as a white mark at the tip of the queen's abdomen. This is the so-called 'mating sign'. Multiple-mating with a number of drones is the rule rather than the exception. The average number of drones achieving successful mating is probable seven or eight, although it may be more. At least one authority says up to twenty! It is not clear how subsequent drones manage to dislodge the mating sign of previous couplings but it is thought that they are pushed aside during the queen/drone struggle.

The mated queen, bearing the last mating sign, returns to the hive and is readily accepted by the workers and groomed. The accumulated load of spermatozoa moves up the spermathecal duct into the spermatheca partly by chemotaxis, a kind of chemical attractant and stimulant, and partly as a result of the motility of the sperms. The process is completed in four to six hours. If she has not received enough sperms on her first mating flight a queen will fly out again, perhaps on following days.

8

The Products of the Hive

Even allowing for contract pollination, the production of honey is the main object of beekeepers, so perhaps it would be useful to try to define what honey is. Nectar is a sugary solution produced by plants and trees from water and nutrients absorbed from the soil and the atmosphere. Partly as a result of the action of photosynthesis and partly by other cellular processes in the plant, sap is converted into a solution of sugars in water which is offered in the nectaries of flowers and sometimes in extra-floral nectaries in the axils of leaves, as happens with beans, cherries and laurels. The nectar contains a high proportion of sucrose (a disaccharide). This is the domestic sugar one buys at the grocers. There are also small quantities of more complex sugars and traces of salts, acids, enzymes and aromatic substances. The water content varies greatly with the plant species. For instance, plum and pear nectar have a low sugar content while it is much higher in many wild plants such as dandelion and charlock. Temperature, humidity and time of the day all have their effect on the secretion of nectar.

Honeybees seek out sources of nectar, collect it and return to the hive with it where it is modified and processed. Much of the water content is evaporated and an enzyme, invertase, is added. This has the property of breaking down the disaccharide sucrose into two monosaccharides, glucose and fructose, in roughly equal proportions. (Glucose is dextrose, and fructose is levulose. The multiplicity of terms is sometimes confusing.)

Honey may therefore be defined as a floral product as extracted from the honeycomb *with nothing added and nothing taken away*.

The nature, quality and flavour of honey is governed by the plant sources from which the nectar came. In a rather confusing way beekeepers refer to 'flower honey' meaning honey other than that derived from heather, although this comes from the flowers of ling (*Calluna vulgaris*). To make confusion worse, they will also refer to 'tree honey'. This is the early season honey from fruit, sycamore, wild cherry, etc.—but lime honey is not usually called 'tree honey'.

Honey is not a simple chemical compound, but a fragile balanced solution of different sugars in water with additions of other substances derived from plants, amino acids, enzymes, aromatic substances and minute traces of elements plus pollen in suspension. It is a very delicate substance that can easily be destroyed by injudicious handling. A good average sample of English honey will consist of:

Table 6.   Contents of Honey

Constituent	Percentage proportion
Fructose (Levulose)	40
Glucose (Dextrose)	35
Sucrose	1
Other sugars	3
Amino acids, vitamins and proteins, trace elements, essential oils, inhibine, pollen	3
Water	18

Wild yeasts are always present, but not in measurable quantities. These will grow and cause fermentation if sufficient water is present and the temperature is 50-80°F (10-26°C). If the water content is lower than 21 per cent the yeast cannot grow.

Honey is extremely hygroscopic, that is, it has a great affinity for water which it will absorb from the air if containers are left open. For example, honey with a water content of 16 per cent will be in equilibrium with a humidity value of 52; at 21 per cent water, equilibrium will be established at humidity of 66; at 29 per cent water equilibrium will not be established until a humidity figure of 76. This is merely another way of saying that honey will always absorb water in normal humidity conditions in temperate zones.

Honey keeps best in air-tight jars or tins and a temperature of 50°F (10°C) or below. It is, in effect a super-saturated solution of sugars and this explains its hygroscopicity and also the fact that most types of honey (there are exceptions) have a natural tendency to crystallize (granulate). Glucose will crystallize more readily than fructose and this is why types of honey with a higher than average proportion of glucose will crystallize unevenly. If this takes place in a glass jar the result is unsightly — a layer of rather coarse crystallization at the bottom of the jar covered by still liquid honey, the fructose portion. Non-beekeepers will sometimes see this and think the honey has 'gone bad' whereas in fact it has only changed its physical, crystalline state.

An extreme example of rapid crystallization is that of oil seed rape honey. This is very high in glucose and as the honey is likely to be entirely uni-floral, crystallization is likely to be very rapid indeed. It will set in the comb and beekeepers who know that their bees are working rape have to extract quickly, before the combs are completely sealed. On the other hand, those kinds of honey which are high in fructose very rarely crystallize at all. The other exception referred to is heather honey. It has a high colloid content and sets in a gel. It cannot be extracted in the usual centrifugal extractor without prior agitation and is usually pressed out of the comb through a straining cloth. Furthermore, the gel is thixotropic—if stirred, the gel will become viscous and pourable, but if left will again set in a gel. It never crystallizes. If heather honey is found to be partly crystallized, and this can be detected by feeling the crystals between the forefinger and thumb or between the teeth, then it certainly is not pure heather honey and has an admixture of other floral honey.

Although fructose and glucose appear to have such different properties, these are entirely physical. Chemically the two sugars are identical, but owing to different molecular structure they have different crystalline and optical properties. The names levulose and dextrose indicate this. Levulose rotates polarized light to the left and dextrose to the right. When crystallization takes place not all the water of solution is taken into the crystals. Some is left behind to dilute the remaining (uncrystallised) honey. This can raise the water content high enough to allow the yeasts to grow and fermentation ensues.

Rapid crystallization results in fine granulation and slow crystallization gives coarse crystals. The optimum temperature for crystallization is about 57°F (14°C) and as the temperature rises, crystallization slows until it ceases at about 94°F (34°C).

Crystallization is initiated round small particles, possibly pollen grains or even air bubbles. Taking these three facts together the beekeeper has in his hands the means of controlling the granulation of his honey. Supposing honey bulk has crystallized coarsely or unevenly (often both conditions go together), if this is heated gently to not more than 100°F (38°C) the coarse crystals will re-dissolve. If the bulk is then 'seeded' with the addition of about 5 per cent of honey that has set in a fine grain, thoroughly stirred and then cooled to 57°F (14°C) it will set in an even, smooth texture.

Commercial honey packers use variations of this heat-seed-cool process to get crystallized honey of consistent set of marketable appearance. The actual method used varies with the sophistication of the plant. Instead of a low heat applied for some time, the honey may be pumped through pipes, flash-heated to a high temperature for a short time and then rapidly cooled, but the same basic principles apply. It is possible to carry the heat treatment further and to force the honey through pressure filters, often of diatomaceous earth. What comes through the filter is a bright fluid with no solid matter suspended. All the pollen remains in the filter. It has a good appearance and a long shelf-life, but is it honey? Judged by our definition I suggest that it is not. The pollen has been removed mechanically and the heat treatment has destroyed some, if not the greater part, of the enzymes and aromatic oils. All heat is bad for honey and excessive heat is excessively bad. I believe too that some flavour is lost in crystallization, but this is unavoidable.

To get the true flavour of pure honey try breaking off a piece of honeycomb taken straight from the hive while still warm. This has a delicacy not to be found in any other form of honey. My own view is that filtering should be confined to straining through nylon mesh. The sort amateur wine-makers use is excellent. This is coarse enough to permit the passage of pollen and fine enough to hold back wax fragments etc. What you then get will win no prizes at Honey Shows but will be a natural product almost exactly the same as the bees left it. Such honey is not difficult to produce, but is more time-consuming jar for jar than commercial methods. The total product, countrywide, coming from careful beekeepers on a small or medium scale and proud of their craft and not entirely concerned with cash, is a useful addition of pure quality to the food resources of the country, and a valuable addition to the home larder.

Nectar Sources
The colour of honey varies from the very pale, almost water-white of willow herb through the greenish colour of lime honey and the dark colour of tree honey, to the red of bell heather and ling. Colour varies with the type of soil and subsoil and may even vary from year to year, which could indicate that weather conditions play their part. Generally speaking the lighter the honey, the milder the flavour.

There used to be a common saying that 'bees follow the sheep'. I am not sure that this is still true in modern farming methods, but it is the case that some of the best honey-producing areas lie in and around the chalky plains and the downs. This is where wild white clover is often abundant and clover needs an alkaline soil if it is to yield nectar well.

Clover has always been, and still is, one of the major nectar plants. The honey appeals to most palates. The flavour is delicate, colour pale yellow to amber and if there is anything like a nectar flow in a clover district the honey is likely to be nearly

100 per cent clover. When crystallized it sets with a smooth fine grain.

Lime is another honey which can be obtained in almost a single floral state. It is a good friend of the beekeeper in some (but not all) years. Density is less than that of clover honey. Limes are liable to honeydew (see p. 123), particularly in hot dry summers, and this darkens and spoils the honey. The honeydew will contain melezitose, a higher sugar, which bees are unable to break down. The nectar of two species of lime, *Tilia petiolaris* and *T. cordata* contains quantities of complex sugars, chiefly mannose and this is abnormally high in hot dry weather. Mannose is a common sugar which can safely be utilized in the human digestive system, but upsets the metabolism of some species of bees. As with melezitose, bees lack the enzyme to break it down into assimilable sugars, so the 'toxic' sugars collect in the blood and thorax. The flight and leg muscles are unable to function and affected bees exhibit the symptoms of paralysis. It is not an uncommon sight in a hot dry season to find masses of bees, many of them bumbles, lying under lime trees dead, paralysed, or apparently 'drunk'. The accumulation of indigestible sugars is usually the reason.

Heather (ling) honey has quite different features from flower honey as we have seen. Its appearance is different too. Air bubbles remaining after extraction cannot rise to the top of the container because the honey has reverted to a gel and these suspended bubbles are characteristic. It is dark in colour and has a strong flavour, not appreciated by some, but very much liked by others. It always commands a higher price than flower honey. The needs of the plants are diametrically opposed to those of clover. They need an acid soil and will not yield nectar on an alkaline soil or subsoil. In some areas ling will grow in a peaty soil overlying a limestone layer. However, they do not do particularly well in these conditions.

Bell heather (Erica cinerea) honey is not gelatinous and can be extracted in the ordinary way in a rotary extractor. The colour is red (almost port wine colour) and the flavour reminiscent of heather. Although often associated with ling, the flowering period is earlier. This will give a chance for a take of bell heather honey before the ling flowers.

Field beans are a good source of nectar early in the year. The honey is mild and pleasant but granulation is rather coarse as is charlock and mustard, both of which will crystallize in the comb as rape does, which we have already noted.

Hawthorn will yield copiously in some seasons, but not always. The honey is dense, usually dark and of a good flavour.

Sycamore will also yield well in most seasons. The honey has a green tinge and the flavour is not, perhaps, one of the best. However, it matures well and any slight unpleasantness soon disappears with keeping. It will crystallize with a coarse grain.

Blackberry is an almost omnipresent plant and a very good friend to beekeepers. It flowers over a fairly long period late in the season, does not seem at all choosy as to soil composition. It is much visited by bees as is its more refined cousin the raspberry. The honey is pale, dense and rather slow to granulate. Flavour is mild and it is a good honey for blending with other strongly-flavoured types of honey.

Sainfoin is worth a mention, although little is grown now. It used to be extensively grown for fodder, but fashions have changed. The honey is golden, has a very fine flavour and crystallizes with a fine grain. One unmistakeable characteristic is the pollen which contains quantities of bright yellow oil. This stains everything within the hive—wax, frames and even the internal walls and the cover board.

The above is no more than a brief look at some of the more important honey plants. The list given in *Appendix 2* may be helpful and for greater detail the reader is referred to Mr F. N. Howes' excellent little book *Plants and Beekeeping* (Faber & Faber, 1979).

Honeydew

Aphids and scale insects live by sucking the phloem sap from plants and trees. Their principal needs are nitrogenous matters and some carbohydrate. They extract what they want and excrete the bulk as a sticky fluid which adheres to the leaves of the host plant. Figures vary greatly with the age and species of the plant tested, but one sample (young field bean) may serve as an example. Analysis of the sap showed a carbohydrate content of 5.8 per cent and nitrogen of 0.24 per cent. The sugary secretion of aphids on the plant had carbohydrate 5.2 per cent and nitrogen 0.11 per cent. In other words, the aphids had retained half the nitrogen but only 10 per cent of the carbohydrate. This sugary solution is honeydew. The carbohydrate fraction consist mainly of dextrose, levulose, trehalose and fructomaltose plus amino acids but, of course, no aromatic oils since the origin was plant sap and the floral nectaries were bypassed.

Bees collect this sweet liquid and process it into what might be called 'honeydew honey'. This is usually very dense, dark in colour (sometimes a rich red), it almost never granulates and has a rather strong, but not unpleasant, malty flavour. Many people like this flavour and indeed it is preferred in Continental Europe where it is highly prized and commands a higher price than floral honey. This is the famous 'forest honey' (*waldhönig*).

Honeydew does *not* come from extra floral nectaries but from plant sap through insect intermediaries. If my definition of honey on page 118 is acceptable then honeydew cannot be classed as honey. There has been much wagging of wise heads at European Economic Community honey group working parties and elsewhere merely on this question of definition. There would be little profit in entering into the argument here so we will try to confine ourselves to known facts as we see them. I strongly suspect that many of the dark types of honey we see in dry hot years do in fact contain a proportion, and sometimes a high proportion, of honeydew. They may be none the worse for this.

It is said that honeydew is not a good winter store for bees because of the high proportion of complex sugars. In some areas of Germany honeydew gathered from larch and spruce can contain an exceptionally high level of melezitose. As I have said, bees are unable to invert this into simple sugars and crystallization is rapid and the set is very hard. During the winter bees cannot dissolve this rock-hard mass. Similar doubts are held about the suitability of heather honey as winter food, this time because of its high colloid content. There is evidence that the safest winter food for bees in pure sugar syrup, provided it is fed early enough in the autumn for the bees to process it.

The Medicinal Properties of Honey

In addition to being an excellent natural food of delicate flavour, honey has medicinal qualities which have been known from earliest times. It is mentioned in most of the sacred books of India, Persia (as it was), China as well as in the Koran, the Talmud and the Bible. Here is one example from the Bible, although there are plenty more; 'My son, eat thou honey, for it is good.' (Proverbs 24:13). Later on Saint Ambrose, the patron saint of beekeepers is reported saying, 'The fruit of the bees is desired of all and is equally sweet to kings and beggars and it is not only pleasing but profitable and helpful, it sweetens their mouthes, cures their wounds and convaies remedies to inward ulcers.' Mohammed said, 'Honey is a medicine for the body and the Koran is medicine for the soul.' One could hardly ask for better recommendations than from Solomon, a Saint and the Prophet!

In antiquity and through the Middle Ages honey was an important medicine and figured in innumerable nostrums and folk cures, sometimes with less pleasant additions. Throat complaints, intestinal disorders, wounds and burns which were common in those violent days were all treated successfully with honey, and still are. There is quite a body of medical opinion that the consumption of honey has a preventive and curative effect in rheumatic ailments quite apart from the well-known belief that bee-stings are an aid and a relief from rheumatic pains.

The rapid assimilation of the simple (invert) sugars comprised in honey through the gut wall without any prior digestion is obviously a source of quick energy, a fact well known to athletes and long distance swimmers who need to maintain their blood sugar levels. Muscles in action consume three and a half times more glycogen than when they are at rest. Glycogen is normally deposited in the muscles, the heart, the blood and particularly the liver, which is a sort of savings bank of glycogen. Normal blood contains about 0.10 per cent glucose. Eating honey increases the blood sugar which is immediately available to replace the glycogen oxidized by muscular effort.

Honey, either by itself or made into an ointment with a drying agent such as white flour has been used as a dressing for infections and wounds through the ages and up to the present time. In recent years the late Mr Michael Bulman, when he was obstetric and gynaecological surgeon to Norwich Hospital, used honey extensively in his post-operative dressings. He was convinced that healing was quicker, more complete and left less scar tissue when he replaced the then more usual flavine and glycerine treatment with honey dressings on gauze replaced daily.

A lady of my acquaintance, when making coffee at a dinner party upset a pot of freshly-made coffee, a good deal of which went over the inside of one thigh. She quickly smothered the scald with liquid honey, pinning a clean tea towel over the lot. She claims that the pain ceased 'in a matter of moments'. In any event, not wishing to disturb her guests, she changed into a clean skirt and rejoined the company making a joke about 'messy and untidy cooks'. No one knew what had happened in the kitchen, not even her husband until bed time. In the morning there was no blister and the red discolouration cleared up in a few days. In contrast, before I kept bees I had a somewhat similar accident with a foot while camping in the Welsh hills. The whole of the top of the foot became one large blister and this did little to enhance my hill walking. Daily treatment with antibiotics by the nearest doctor was effective and I was truly grateful for his care and attention. However progress was slow. The wound was not completely healed after four weeks of treatment.

There is no doubt that honey is bactericidal and I think this is due to two main reasons. Because of its strong hygroscopicity honey absorbs water from any available source with which it comes into contact. Like any other living organism bacteria need moisture to maintain life and multiply. Honey draws the water from these unicellular forms of life and they die. Secondly the inhibine listed as an inclusion in honey is, I think, due to an enzyme added by the bees. This initiates production of peroxide so that honey contains a positive antibacterial substance. The advantage of this to wound dressings needs no emphasis.

The Microscopical Examination of Honey

Melissopalynology is the scientific term for the examination of the solid particles suspended in honey and the interpretation of the findings. Apart from the fascinating world opened up by the beautiful designs of pollen grains, there are two main reasons for the study.

There are a few types of honey which come from a single floral source; for example, heather, clover, rape and lime (sometimes). Nectar flows in these cases are either concentrated or come from the only flowers available to the bees at the time of foraging so that it is a simple matter of field observation to confirm the source. However, many beekeepers only take their honey crops off the hives once or perhaps twice a year. During the active season the bees have been moving the nectar and half-processed honey about within the hive so that the final product is a real mixture. It has been said with reason that English honey owes its appeal to its multi-floral origin. It is interesting to find out where the bees have been and what the main floral ingredients of the honey are. Pollen analysis is a quick way to find out. Suspect honey can be examined too for the undesirable inclusion of soot particles, fungal spores etc.

The second important reason is the determination of the geographical origin of the honey. Every pollen grain is quite distinctive and recognizable. As Dr Louveaux has said, 'Honey carries within itself its own certificate of origin.'

The flora of the United Kingdom differs from that of, say, Spain or Hungary. Examination of the pollen will show where the honey has come from. It is not only what is there that is important. What is missing is equally important in identification. The microscopist/beekeeper can, with a little practice, recognize the pollens common in his own area and be on the look out for strangers.

The price of good English honey is at a premium as compared with that of foreign origin. Unhappily there have been too many cases of foreign honies masquerading under misleading 'English' labels. Pollen analysis of these types of honey will disclose their geographical origins unerringly.

The ultimate test is in the Courts, where only findings of Public Analysts are acceptable. Those of experienced beekeepers are not. Unfortunately, very very few analysts are in a position to give any kind of credible opinion as to the floral or geographical origin of a given sample of honey. I do not wish for a moment to impute their technical ability. It is obvious that this far surpasses that of amateurs. The difficulty lies in the identification of pollen grains and the interpretation of results. So far there is no published key of pollen grains which is accepted as the standard work on the subject. The fact that many bee-keepers, by training and experience are perfectly capable of analysing a sample and of giving a reasoned opinion as to its origin has no legal validity.

For the ordinary beekeeper not concerned with fine legal points, but interested enough to experiment for himself, the only essential (and expensive) piece of apparatus is a compound microscope giving a magnification of at least 400 diameters. Many good instruments exist up and down the country in family cupboards, collecting dust on Victorian memorabilia, in private use among keen beekeepers and amateur entomologists, in schools and technical colleges and in Beekeeping Associations in boxes under the Secretary's desk.

Preparing a Slide

The technique of preparing micro-slides is basically simple and consists of concentrating the solid particles from a weighed sample of honey, staining them and mounting them on a glass slip for examination at leisure.

My own method is as follows:

1. Stir the honey. Warm it a little. If it is crystallized get it completely liquified.
2. Weigh out 10g. A 50ml beaker is a suitable container.
3. Add warm water up to 25ml and stir until completely dissolved.
4. Pour into centrifuge tubes. 15ml size is the best.
5. Balance tubes.
6. Centrifuge for 5 minutes.
7. Decant. Get sediment into one tube and refill both tubes with clean water. Balance tubes.
8. Centrifuge for 5 minutes.
9. Decant. Suck up sediment with a fine pipette and spread on a 3 by 1-inch (about 7.5 by 2.5cm) glass slide.
10. Dry on warm plate. (This could be the warm, not hot, side of the top of a domestic boiler.)

11. Add a drop of stained glycerine jelly. This can be bought at Naturalist and Microscopists stores. The standard article is usually a bit too heavily stained for my liking, so I buy a tube of clear jelly with every tube of stained and mix the two together.
12. Add cover glass.
13. Replace on hot plate for 15 minutes.
14. Allow to cool (some hours) and when quite cold next day clean off any surplus jelly with cold water.

The Centrifuge

The laboratory worker weighs and balances his centrifuge tubes by using a chemical balance. This is beyond the reach of most of us. With regard to Step 2, a perfectly adequate substitute for weighing can be bought at winemakers shops. It consists of a dish holding about 50ml attached to one end of a balance arm, the other end having a counter-poise weight. The balance arm is graduated in ounces and grams and slides in a rocker fitted with a bubble level. The principle is that of the steel yard and the device is surprisingly accurate.

Figure 21. The head and tubes of a home-made centrifuge.

It is essential to balance centrifuge tubes precisely. I use an old apothecaries pocket balance which I have adapted by removing the weight pans and substituting loops, each made of one strand of picture wire, which hold the tubes. It is so sensitive that one small drop of water is enough to upset equilibrium. The main fulcrums are steel knife edges and I am sure a similar arrangement could easily be made with the working parts suspended on razor blades. A U-shaped piece of copper wire could be used to zero the set up and the wire squeezed on tight when level is achieved.

Not everyone possesses a centrifuge. For some years before I bought myself the proper article I improvized one using an ordinary electric drill clamped firmly in a bench drill. The incorporation of a dimmer switch in the power cable gave control of speed from very slow to maximum. The centrifuge

Figure 22.   The home-made centrifuge attached to an electric drill, gripped in a vice.

tube-holders were short lengths of copper water pipe close at one end and burred over at the other to form lips. The copper pipes were allowed to swivel in a channel made from aluminium sheet. My pivot was a small mandrel—the kind that is sold for fitting a grinding wheel to a hand drill. This needs to be short and rigid. Any 'whip' makes the whole thing uncontrollable. If any reader should decide to make a similar device may I please emphasize as strongly as I can that this is potentially a most dangerous piece of apparatus. It must never, repeat NEVER, be switched on unless it is completely shielded. I made a wooden box which fitted over the whole thing and was screwed to the bench. Access was by way of a hinged lid. This was cumbersome but safe. At the speed of revolution obtained, any flying bits which might break off become lethal projectiles.

Figure 23. Centrifuge safety cover. This should *always* be used when the centrifuge is in operation.

Sedimentation—An Alternative Method

This method is effective but has some disadvantages.

Make up a solution of honey in water as in Steps 1 to 3, but instead of pouring it into centrifuge tubes as in Step 4, pour it into a tall test tube. The taller the better, providing you have a pipette that will reach the bottom. Lighter pollen grains will sink slower than the heavier grains so you will get a fair deposit in twelve hours, but longer than this will give a more representative spectrum of the pollens. Decant the water off. If this is done in one smooth movement all but a few drops will pour out leaving the sediment undisturbed. If clean water is added, the tube agitated and then allowed to settle for another twelve hours (twenty-four is better) most of the honey and yeasts will be eliminated. This step can be repeated a second or even a third time if a really clean pollen sample if needed. Transfer this to the centre of a micro-slide with a pipette as in Step 7 and proceed with Steps 10 to 14.

By using a centrifuge much time is saved, a complete deposit of all suspended solid matter is obtained so that a quantitative analysis of every type of pollen is possible, and by repeating Step 7 it is possible to wash out all the honey.

With sedimentation no special apparatus is needed and the method can be carried out anywhere. The two disadvantages I have found is that some honey is always carried over to the micro-slide and also, in warm conditions, the longer immersion in water allows the wild yeasts to develop with the result that slides sometimes have a background of thousands of yeast cells. However, these do not take the stain and although the slides may look unsightly, the stained pollen grains stand out well.

Learning to Identify Pollen

Now for the hard part—identification. There is no quick way to gain expertise in this, but recognition of commonly occurring grains comes quickly with practice.

It is a good idea to build up one's own collection of type slides. During the year, as local flora comes into full bloom, take pollen directly from the anthers by shaking or scraping them on to a micro-slide. Use mature anthers as far as possible, otherwise you may get immature and possibly distorted grains.

Many pollens have a high fat or wax content and this can be removed by adding a drop of ether, alcohol or acetone (try nail varnish remover). This will spread the grains and wash out some of the oil globules. Carefully wipe round the edge of the spread ring with a tissue when it is nearly dry. Now follow Steps 11 to 14 in our schedule.

There is much to be said for building up your own collection of pollen grains taken from the most important floral sources in your area. There is no reason why you should not add the more unusual or exotic pollens you may come across in your garden or your neighbours or indeed on your travels. However, the important thing is to get a named collection of pollens from flowers which occur frequently within flying distance of your apiary. In time this will provide you with a standard of reference to which you can turn in case of need. The tidy mind will find no difficulty in devising a simple system of cross reference for his collection for ease of identification.

Why should you wish to identify pollen at all?

I think there are four main reasons.

1. By examining and identifying the contents of the corbiculae of pollen foragers the source can be found.
2. Nectar gatherers inevitably get dusted with pollen from the flowers they visit. Examination of the pollen scraped from the body hairs will indicate the flower species visited for nectar.
3. The examination of pollen taken from dead or dying bees suspected of spray poisoning will again indicate the crop at fault.
4. Identification of the pollens in honey will indicate the local or foreign origin of the honey in question.

There is inadequate literature for the identification of pollen and a collection from known species like this would be an invaluable help for individuals and Beekeeping Associations. One very useful book is Dorothy Hodges' excellent *Pollen Loads of the Honeybee* (see *Further Reading*). Apart from colour charts of the pollen loads to be seen in the corbiculae of returning foragers there are line drawings of 160 types of United Kingdom pollens. One word of warning is needed about these. Mrs

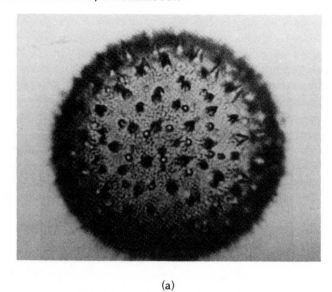

(a)

(b)

Figure 24.   Pollen grains are identified by the configuration of the outer shells. The two examples here are (a) mallow and (b) ivy.

Hodges drew them from life under the microscope and, of course, reduced three dimensional objects to two dimensional drawings of quite exceptional beauty and clarity. It is unlikely that the beginner will see the detail she has delineated and certainly not immediately. As with any other microscopical examination, the fine focusing control will need to be constantly racked up and down in order to get a complete mental picture of the objects on the slide. This is common to all microscopical examinations. The microscope projects an optical section and a complete view can be built up by looking at a series of such optical sections. The higher the magnification the shallower is the depth of the section.

Beeswax

It is interesting to reflect that bees convert part of their food into a substance which they can use to build the combs which form their nest and in which they take shelter from the elements, rear their young and store food for the winter. It is a very poor conductor of heat so has a considerable insulating effect. Perhaps this fact makes them unique. Other animals use material they obtain from their environment to construct their nests, for instance, the wasps beautiful galleried nests are made from wood and similar material chewed by the worker wasps— but I think the bee is the only social animal to manufacture the building material within her own body.

On the underside of the abdomen the worker bee bears four plates, half of each being a very thin membrane (the so-called 'wax-mirrors'). Behind each wax mirror lies a set of wax glands which become active at a certain stage of the bee's development and when wax is needed by the colony. The bees feed heavily on nectar, honey or sugar syrup according to what is available and hang in festoons. The wax mirrors are permeable and the oily liquid produced by the wax glands flows through the wax mirrors and solidifies into tiny white flakes of wax on contact with the air. The legs pass the wax flakes forward to the mandibles where they are chewed and mixed with other glandular secretions to become the small particles of wax which are the building bricks of the combs. Although originally white the wax acquires the familiar yellow colour by the inclusion of pollen and particles of propolis.

The wax has a characteristic odour and a melting point of about 145°F (63°C). It is insoluble in water and alcohol but completely soluble in cold chloroform and ether and in benzene and carbon disulphide at temperatures of 86°F (30°C).

Blocks of cast pure beeswax will develop a bloom, particularly in cold weather. This is sometimes thought to be due to a mould or mildew but in fact is neither. It is due to a re-arrangement of the surface molecules of the wax and may be taken as an indication of purity. It does not seem to occur with adulterated wax.

Beeswax is extensively used in the cosmetics industry and, when bleached, is the basis of most cold creams, lipsticks etc. It was once used in the electrical industry in the manufacture of capacitors, but has been superseded in recent years by cheaper waxes I believe. It has always been used in the manufacture of fine candles and in antiquity the Church insisted on the purest source of light for religious ceremonies. It required that candles of pure beeswax be burnt at the altar during Mass or the Benediction of the Blessed Sacrament. It is still used for the best church candles, but the residual smell after sniffing sometimes compels one to wonder whether perhaps there is sometimes an adulteration of cheaper waxes.

The wax is used in many industries and for many purposes: in the pharmaceutical industry—salves and ointments; in the dental trade—impression wax, but now replaced largely by hydro-carbon based materials; metal casting; polishes for furniture and leather. There are too, a host or minor uses: waxing linen thread for the best hand leather sewing; crayons; chewing gum; waxing archers bow strings; polishing lenses. Finally, on our own humble level, beeswax is needed for the making of wax foundation for hive frames, either made at home or professionally.

Enough has been said to indicate that beeswax is an important and high-value by-product of the hive, and not a scrap of it should be wasted or thrown away. Old comb and odd bits of brace comb should all be saved for rendering down later. This is most conveniently done in a solar wax extractor, another useful piece of equipment easily made by the handyman. In addition, there are commercially-made wax extractors utilizing steam or

Figure 25.   A solar extractor.

hot water. It seems practical, however, to use the natural energy of the sun. In a recent spell of sunny weather in June shade temperature in my garden was 75°F (24°C). In the solar extractor it was 210°F (just below 100°C). Is this a further proof of the utility of solar heating panels?

Propolis

Propolis is an aromatic resinous substance collected by bees from trees and plants such as poplars, horse chestnuts, laurels and pines. The name derives from two Greek words, 'pro' meaning 'before' and 'polis' meaning a 'city', because in ancient times it was noticed that certain races of bees had a habit of building walls of the substance just inside the entrance to a hive and in front of the combs.

Colour varies from yellowish-red to dark brown with a green tinge. In warm weather it is sticky and pliable, but in cool conditions it hardens to a brittle consistency. In the raw state the taste is rather bitter and 'hot' but not unpleasant.

Bees use propolis to seal any cracks or crevices in the hive too

Figure 26.   Propolis. Fairly large lumps like this are more saleable than dusty scrapings.

small for them to pass through and to fasten movable parts such as frames to their supports. They also use it as a varnish on the inside walls of hives and apply a thin coating to the walls of cells when it gives enormous added strength. When old, blackened combs are melted down in a solar wax extractor, in addition to the accumulated cocoons of successive generations of brood, a considerable amount of dark propolis will be left behind and some will pass through in the molten wax itself. This can be removed when the wax is again melted. The propolis sinks to the bottom and the clean wax can be decanted.

The bees will use propolis to encase and seal in any unwanted object in the hive too large or heavy to be dragged out. Many a beekeeper has found on the floor of a hive a dead mouse completely covered with propolis and stuck to the floor so that the hive has remained clean and sweet-smelling. In addition to the physical sealing of the mouse's body, the anti-bacterial properties of the propolis prevent putrefaction.

Beekeepers have known of the existence of propolis for as long as there have been beekeepers, but I think the first written reference to it was made by Anton Jansha in 1776 in his book *Complete Instructions for Beekeeping* in which he described its use by the bees and called it 'putty-wax'. It is said to have been a constituent of the varnishes applied to Amati and Stradivarius violins and 'cellos in Cremona and to influence the resonance of the instruments'.

Propolis foragers bite into a source they have found and tear off lumps of it by bracing themselves backwards with their front legs in exactly the same way as predatory birds tear lumps of flesh off their prey. The lump of propolis is passed backwards via the front and middle legs and finally into the corbiculae on the hind legs and there patted into shape by the middle legs. It may take anything from a quarter of an hour to an hour to get a full load.

Back at the hive the propolis is removed by other workers and used at once where it is needed, perhaps with a little wax added. Propolis is never stored as honey and pollen is so that the propolis forager remains in the hive until all her load is taken from her. She does not appear to do the cementing herself but leaves this to other workers.

There is a considerable body of evidence that the antibiotic properties of propolis are of use in relieving many human complaints and illnesses. The pharmaceutical industry produces lozenges, capsules, creams and tinctures based on, or containing, propolis. These are pleasant ways of taking it but the beekeeper, with access to fresh raw propolis, can taste it if he or she so wishes and test its effect by sucking a couple of chips. I have repeatedly done this myself and believe it has good effects, especially in such conditions as sore throat. I also have a notion that it improves circulation and slightly raises body temperature.

Pollen

In depreciating the fine filtration of honey which removes the pollen grains we have noted briefly that pollen has considerable nutritive value. To expand this it is worth noting that it contains twenty-one amino acids (the constituents of its protein content) and at least seven vitamins, fourteen minerals and twelve enzymes.

There is a sale for pollen for dietary supplement. The pollen is obtained by placing a pollen trap over a hive entrance. These consist basically of a grid through which the bees have to pass. This is of a size which enables free ingress and egress to the bees but which is fine enough to knock off the pollen loads from the legs of pollen foragers. The pollen falls into suitably positioned trays. The pollen trap must not be left on the hive indefinitely as this would deprive the colony of its essential protein and prevent the raising of brood. Normally pollen traps are left on the hive for only a few hours each day.

Pollen has been used as a method of determining the intake by plants of heavy metals in different kinds of soil.

Royal Jelly

Royal jelly is the substance produced in the hypopharyngeal glands of young worker bees and fed by them to worker larvae for the first few days after hatching and to queen larvae for the whole of their larval life. It may be that there are hormonal differences between the brood food fed to young worker larvae and that fed to queen larvae, quite apart from the possible addition of honey and pollen to the diet of worker larvae. Royal jelly is highly nitrogenous and is the food supplied to queens for the whole of their egg-laying life.

When taken by man it is claimed that it has beneficial effects on eye malfunctions and on general health. It seems to be a clearly established fact that it induces a state of euphoria, but this may be a psychosomatic effect. More research is needed before definite and valid conclusions can be drawn.

Essentially the production of royal jelly is effected by keeping colonies in a queenless state for comparatively long periods and robbing them of the royal jelly they deposit in the resultant queen cells. It is not a process which commends itself to many beekeepers.

Venom

Bee venom has been used in recent research into alleviation of rheumatic illnesses. There is reason to believe that the belief that bee stings relieve rheumatic pains is more than just folklore, but may be due to bodily reactions which are not clearly understood yet.

There is current investigation into the possibility of inducing immunity from bee stings, at least to a partial degree, in people who have an allergy to bee stings. The old treatment used to be a course of injections of extracts of whole bees. In many cases this proved to be of doubtful effectiveness. The new method under trial is a course of injections under medical supervision of pure venom starting in a very dilute form and gradually increasing to full strength eventually. Preliminary results are encouraging.

The method of collection is interesting. We know that a bee cannot withdraw her sting when she implants it into the flesh of a mammal and that she tears out her final abdominal segments in trying to free herself. The problem was to find a method of milking the bees of their venom without killing them. So far, the best way of doing this is to stretch a very thin membrane such as silicone rubber over a metal grid. This is placed on a strong colony and a low charge of electricity put on the grid. This annoys the bees without harming them and they promptly sting the source of their annoyance. The stings protrude through the membrane and the venom can be scraped off. If the grid and membrane is below the colony the venom falls into a tray. In either case the membrane is so thin that the bees can withdraw their stings with no difficulty.

9

Swarming and Increase

A ll living organisms must reproduce their kind if the species is to survive. This is so important a part of the life cycle that natural selection has ensured that the urge to reproduce is a very powerful one indeed.

In the case of the honeybee, reproduction has a twofold aspect. In the brood nest at the height of the season scores of young bees emerge every day to replace their older sisters whose summer occupations have literally worn them to death. During the period of honey-flows, the life-span of a worker from egg to death may be no more than six to seven weeks while bees hatched from eggs laid in autumn survive into the spring. The controlling factor is the amount of work each worker has to do. This is individual reproduction in order to keep the colony at optimum strength. As we have seen, bees are social creatures and the colony is the unit of survival and must reproduce itself if the species is to survive.

The honeybee has evolved a system by which colonies of bees which are well-founded split into a number of smaller units, each headed by a queen, but at the same time with a controlling mechanism in operation to ensure that the colony division does not continue to point where there are a large number of units too small to survive. This mechanism also ensures that the divisions take place in a time of the year that is most likely to provide the weather conditions for proper establishment of a new home. This is what we call swarming.

In describing swarming, or any other behavioural pattern for that matter, I do not wish to imply that bees make conscious decisions as to courses of action to be followed. I believe that what they do springs from reactions to stimuli they receive, either externally or internally, conditioned by their own neural make-up and that very little of what we call 'reason' comes into it. This may not be altogether a bad thing. We make our decisions on grounds of logic or reason (or claim that we do) and seem to make a pretty mess of it. Our bees, acting on instinct, seem to manage things rather better.

Origins of Swarming

There are two main theories of how swarming arose. One is that swarming has developed from the alleged habit of a group of bees accompanying a virgin queen on her mating flight. The second is that swarming is an offshoot of the habit, still common in some species of Eastern bees, where the whole colony decamps to a new nesting site. Sometimes this is the result of deteriorating conditions, but sometimes there is no obvious reason for the departure.

Effects of Queen Substance

When a young worker emerges from her cell she is already equipped with all the skills she will ever need and with a set of inherited behaviour patterns which we call 'instincts'. One of these is to build queen cells unless prevented from so doing. If there were no checking device on this instinct queen cells would have been built by the score in colonies and the race of honeybees would have died out aeons ago.

We have seen that the queen is constantly groomed and licked by her court and that in doing this her attendants get small doses of an aromatic exudation from glands in the queen's head. This spreads all over her body. It is a complex substance but its two main constituents, 9-hydroxydecenoic acid and 9-oxodecenoic acid, have an odour which is very attractive to workers so it is eagerly sought after. We have also seen that bees are constantly exchanging food; anything absorbed by a group of bees finds its way quickly throughout the colony. One effect of queen substance is to inhibit the bees' natural instinct to build queen cells. The foragers, being away from the hive more than the house bees, are likely to get a smaller ration of queen substance and the house bees proportionately more. This must be effective in increasing the efficiency of the scheme.

The distribution system just described is, of course, an over-simplification and much more needs to be found out. For instance, we do not know how the saliva of the workers affect the queen substance before it is passed on. Whatever the whole story turns out to be, it seems certain that the influence of a mated fertile queen on her colony is initially chemical. This is not such a revolutionary notion as might appear at first sight. Minute doses of chemicals or drugs, whether introduced into the body from its own endocrine glands or from external sources have enormous effects on large animals, as in man. It is not surprising that dramatic effects are apparent in small animals such as bees.

The queen substance theory may explain the building of queen cells and the raising of queens prior to swarming but it does not explain satisfactorily the initiation of the swarming impulse. A number of possible explanations have been put forward. Some are theories of the basic underlying causes and others are suggestions of conditions likely to influence the onset of swarming.

The typical swarm is a good cross-section of the hive population—some drones, workers of all ages, except the very young, plus a queen. The queen does not lead the swarm. Indeed, she is reluctant to leave the hive and in many cases is physically shoved out by the workers. The swarm usually clusters on some support not too far away from the hive until scout bees have found a new nesting site (or a beekeeper has 'taken' the swarm). During this time the swarm is very vulnerable to predators and deteriorating weather conditions.

The Brood Food Theory (Gerstung's Theory)

This theory was very popular around the turn of the century but has become rather devalued as a result of more accurate information on the bees' life cycle, showing that the theory is based on a fallacy. The basic belief is that the division of labour among workers is rigid and solely dependant on the workers age, and this has been shown to be untrue. After the normal spring build-up of population, the queen's rate of egg-laying levels off and the stage is reached when there are more nurse bees than are required to feed larvae. The brood food glands of the nurse bees continue to develop and as they cannot get rid of the brood food by feeding larvae they turn to building queen cells instead. We now know that bees produce brood food according to need and controlled experiments have shown that a surplus of nurse bees does not, by itself, lead to queen-rearing.

Congestion

A hive which is not big enough for the colony which inhabits it will almost invariably lead to swarming. This is not simply because the queen has insufficient room in which to lay. There must also be ample room for stores and the processing of nectar and room for 'parking' too. When most of the bees are at home there is a need for space for them to rest, and to move around in. In spite of the 'busy bee' notion, bees spend a good deal of their time in the hive doing nothing—apart from nectar-flow times of course. It seems that restriction of space for the queen to lay in is far less important a factor in swarming than congestion in other parts of the hive.

Wax-making

It is said that bees of wax-making age need to build comb and if there is no opportunity for this then swarming is more likely. Beekeepers make a habit of regularly removing a couple of frames or so and replacing them with frames of foundation for this reason. Comb renewal is certainly good for hive hygiene but I doubt if it has any effect on swarming except by reducing congestion—two frames of foundation give more space than two fully drawn combs.

Effects of the Season

A favourite topic of conversation in beekeeping circles hinges round whether it is a swarming or a non-swarming season. It is a fact that the incidence of swarming, in general terms, varies greatly from year to year and from district to district. I do not think that any correlation between this and weather conditions has been satisfactorily established and the causes are obscure. Perhaps a season when colonies build up rapidly in the spring earlier than usual is likely to result in a spate of swarming or that a year of little swarming is likely to be a year of light honey harvests. The truth is that we do not know what it is that sets off a season of numerous swarms or what the correlation is between weather and swarming is, or if one exists at all.

Strain of Bees

Some strains of bees are well known to be inveterate swarmers and others swarm only very occasionally. It is fair to say that most bees in this country are mongrels (or hybrids, if the word is preferable). Because of hybridization many characteristics crop up and these can be partially bred in or out by selective breeding. Many good beekeepers have done just this. By culling they have eliminated, or at least reduced, undesirable characteristics and have increased from those strains which have shown desirable traits. The tendency to swarm excessively or lightly is one such character. I cannot help feeling that culling poor and unproductive stocks and breeding from the best is the right way to improve the quality of bees anywhere and is much to be preferred to buying queens of foreign origin with wonderful reputation in their country of birth. In any event, current legislation is putting a much stricter control on imports of bees than in the past. Culling should be a continuous process. Because of the cross-bred nature of our bees, unwanted traits are bound to crop up from time to time and these should be watched for.

Sunlight

Colonies sited in open positions so that the hives are in constant sunlight are, on the whole, more prone to swarm than those in the shade, so it is prudent to site hives where they can get some shade in the hottest part of the day.

Age of Queen

Young queens are less likely to swarm than those older. A policy of regular queen replacement will therefore tend to reduce swarming (but not necessarily obviate it altogether).

Other Considerations

It has been said that many drones in a colony will lead to swarming. This is a chicken and egg situation. Although it is true that no swarm will issue without its complement of drones, there is no evidence that the number of drones in a hive triggers off swarming. It is said too—too often to be ignored—that a good nectar-flow will prevent swarming as long as the flow lasts. Weather also has its effect. Swarms come out in fine warm weather and stay in when it is cold and wet.

It is generally true that large colonies will swarm more readily than small.

Will frequent manipulations provoke swarming? It is true that when colonies are frequently and regularly examined more swarming preparations are seen than where bees are, on the whole, left to themselves. This is simply a result of 'the more you look, the more you see'. Preparations for swarming go on in hives much more frequently than is popularly supposed. Of the swarms which come out and are seen, almost an equal number are started but the queen cells are destroyed by the workers for one reason or another and the swarm is aborted. A swarm only takes about ten minutes to come out and decamp to a clustering place and many swarms remain undetected as a result.

Swarms and Swarm Control

Dealing with swarms and with colonies preparing to swarm is time-consuming and in a 'bad swarming' year can be frustrating. It is a rock on which may a beekeeping venture has foundered. Many people now keep their bees in urban or suburban areas and this, I think, raises a social duty. I feel strongly that we have a moral obligation to keep our bees so that they are no nuisance at all to our neighbours. You and I may enjoy galloping around the neighbourhood after errant swarms explaining as we go that the bees are very good-tempered really if not interfered with. Mrs Smith next door may take a different view. On a warm

Sunday afternoon (unwanted swarms always seem to come out on Sunday afternoons) when she is having tea in the garden with her in-laws, the appearance of an enormous cloud of whirling insects each armed with a deadly weapon is less than welcome and does little to forward neighbourly relations.

When a swarm issues, the work-force of the colony is split into two units. The parent stock is considerably weakened and seems to mark time until a new queen is mated and laying. It is true that a swarm always works with a will, but it has a great deal of comb-building to do. If it is given drawn foundation and there is anything like a nectar-flow the combs can be filled rapidly. But then comes a pause and new brood has to be raised and frequently a swarm when established as a new colony will replace the old queen by supersedure.

Excessive swarming then is not to the advantage of the bee-keeper and can create social problems. To the large-scale beekeeper it represents a loss of potentially productive stock and is time-consuming. How should we prevent it? We have seen that swarming is a natural process so I think we can assume that total prevention is not possible, and even if it were, it would be undesirable. The beekeeper who says he has colonies which have not swarmed for ten years is probably deluding himself. He is likely, and lucky, to have a strain of bee which regularly supersedes—and after all this is only a kind of internal swarming. The amount of sperm a queen receives at a good multiple mating is unlikely to last for more than four years or so and her production of queen substance will be falling off before that.

Swarms can be lost. The best of beekeepers can often miss a swarm, even in a home apiary. A swarm will hang quietly with few fliers after its initial settling and will remain so until the scouts have found a new home.

If we cannot prevent swarming altogether, we can try to avoid those things which we have seen are conducive to swarming. The more important points are:

1. When increasing stocks, use as breeder colonies those with a good record of minimal swarming.
2. Be generous with ventilation.

3. Provide ample space for expansion in the brood chamber and in the supers for honey storage and processing.
4. If at all possible site hives so that they get some shade in the middle of the day.
5. Plan a programme of regular requeening with young queens.

In spite of your best endeavours colonies will decide to swarm from time to time. The only thing to do is to make regular inspections during the months when swarms are likely—to be safe, at weekly intervals—and to have a plan to put into action if preparations for swarming are seen. In doing so you will be dealing with the situation before the swarm issues. This is what beekeepers mean by 'swarm control'.

Many systems of swarm control have been devised. Mostly they consist of destroying queen cells, separating the queen from the brood and giving her more laying space or variations of all three.

The oldest plan of all was to let the swarm issue, catch it, kill the old queen in it and return the swarm to the parent hive. There is a danger that the swarm will come out again, even larger than before, with the first virgin to hatch, so to avoid this it was the practice to go through the colony and destroy all queen cells but one. This creates another hazard. The virgin may be lost on her mating flight or bad weather may prevent a mating flight. If she does not mate within three weeks of emergence successful mating is unlikely. This system is not swarm control at all, but merely consists of allowing the colony to swarm and then dealing with the swarm afterwards.

The Pagden method too is a way of dealing with a swarm after it has issued. The system was originally published in 1870 when fixed comb hives were in vogue, but it can be practised equally well with modern movable frame hives. The swarm is caught and hived. The new hive complete with swarm is placed on the site of the parent colony which is moved to a new site some way away. The swarm is reinforced by the foragers from the parent colony and from the old hive, if all goes well, a virgin emerges and is mated. It has been so depleted (swarm plus flying bees) that no casts or after-swarms are likely. A Demaree (see page 152) variation of the above is to hive the swarm and place it on the old

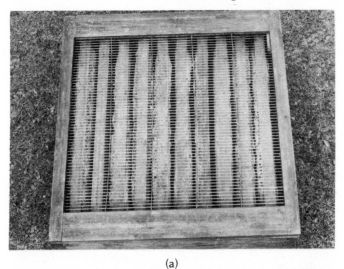

(a)

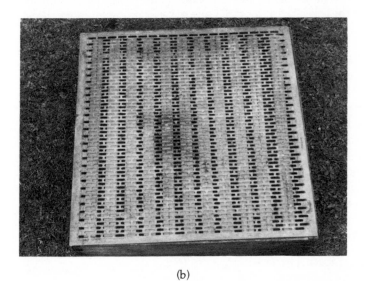

(b)

Figure 27. (a) A wire queen excluder, framed. (b) Slotted zinc excluder, unframed. Square hives, e.g. British National, will take these excluders with the wire or slots, either across the frame or parallel to them, as desired.

site. Fit a queen excluder on this followed by any supers then another queen excluder and finally the old brood chamber with all the queen cells destroyed. The bees will build more queen cells, so it must be gone through again in not later than seven days and any fresh queen cells destroyed. If the system works the number of colonies will not have been increased, it will have been requeened with a young queen and no foragers will have been lost.

The Demaree system can be used when queen cells have been started or as a routine system of control. The brood boxes are carefully examined frame by frame and any queen cells which may have been started are destroyed. The combs are then sorted. An empty brood box is placed on the floor board and in this is placed one comb of unsealed brood, eggs and the queen. The remaining space is filled with empty combs. A queen excluder is placed over this, then the supers followed by a box containing the rest of the brood. The bees may raise queen cells in the top box so this will have to be examined in seven days and any queen cells found must be destroyed.

The colony must be strong enough to look after the brood in the top box and at the same time expand in the bottom box with the queen.

Poor weather following a 'Demaree' can give disastrous results. It is claimed that the process can be repeated through the season. It should be remembered that the system was devised in America where nectar flows tend to be heavy and prolonged and where fine weather over a long period can be expected. A less drastic plan is sometimes practised in the United Kingdom which consists of a half-Demaree. The queen and about half the brood are left in the bottom chamber which is completed with empty combs and the rest of the brood moved to the top box. This should be the youngest brood. In practice this is usually carried out when a double brood-box method of management is used by sorting the combs out into two lots—youngest and oldest in the two boxes. In this case it is not essential to find the queen. If, after the bees have settled down, she is found to be in the top box, this and the bottom box are interchanged. Any queen cells found in the queenless box are destroyed, at the same time seeing that there are most empty combs below and all food combs above as far as this is possible.

The Snelgrove Method

In 1932 L. E. Snelgrove devised a variation of the Demaree system. A screen board is placed between the supers and the top brood box instead of the second queen excluder described above. This screen board fits over the boxes exactly as an inner cover would, but has the central hole permanently covered with wire gauze of sufficiently coarse guage so that the bees can communicate through it and the passage of warm scented air is unimpeded. The rims round the edges need to be a little thicker than normal—say $\frac{3}{8}$ inch (9mm). On three sides wedges of about two inches (50mm) in pairs are cut in the rims one above the

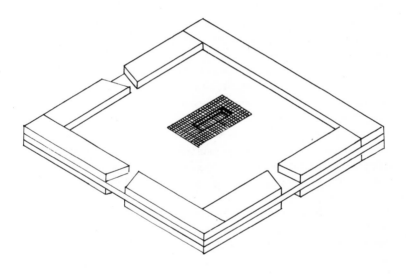

Figure 28. Snelgrove screen board. Triangular wedges are cut from three of the wooden rims both top and bottom. The wedges are retained for inserting and removing as described in the text. An improvement would be to have 'captive' swivelling pieces instead of triangular wedges.

other. The wedges are made so that they can be removed or, better still, secured by one countersunk screw so that they can be swivelled open or shut as desired. The board is placed on the hive with the plain side (i.e. with no wedges) towards the front.

The top wedge on one side is removed so that the bees can fly from the top box. After five days this wedge is replaced and the wedge below it is removed and at the same time a top wedge on the opposite side is removed. Flying bees will emerge from the newly-opened wedge in the top box and on return will come back to the side they have been using, go through the bottom opening and join the foragers in the supers and the bottom brood box. Because of the gauze in the screen board they will have the same odour as the lower boxes and no fighting will ensue. At five-day intervals the operation is repeated with the other pairs of wedges and as a result a strong force of foragers will be built up in the supers and the bottom box and the top box will be populated by young bees which will be easy and pleasant to handle.

Queen cells will be built in the top box and three courses are now open. All the queen cells can be destroyed, the top box left where it is and the screen board removed. The top box then becomes a honey super. Alternatively, all queen cells except one can be destroyed and a queen allowed to emerge and mate. In this case *all* the bottom wedges must be kept closed and one top wedge left open, preferably the one at the back. The newly-mated queen can be used to requeen the existing colony or held in reserve for use when needed.

Finally, the combs in the top box can be divided into two or possibly three nuclei, each with a ripe queen cell and left on the colony with all bottom wedges closed. Each nucleus should have its own entrance/exit and they should be separated by well-fitting division boards and separate covers or quilts. With luck the queens should mate and provide three nuclei headed by young laying queens.

One danger of the scheme is that, should a spell of bad weather intervene half-way through the operation, it is difficult to feed the bees in the bottom box should this be necessary. It is rather fiddly and the time intervals must be carefully adhered to. Nevertheless it is a very useful exercise for the beekeeper who has passed the beginners stage and a lot may be learned about bees and beekeeping from it.

The Artificial Swarm

As the title suggests, this is a method of passing a colony through the motions of swarming but under controlled conditions and, as in the natural swarm, separating the queen from the brood. There are a number of variations, but perhaps the basic, and most satisfactory way is to divide the colony so that the queen-right portion is left on the original site without any brood and all the brood is removed to another position.

To do this, find the queen and place her and 'the frame on which she is found in an empty brood-box on the original stand and fill up with frames of foundation. If any frames of drawn foundation are available this will help to give the bees a good start instead of having to draw out a whole box of foundation. Replace the queen excluder and the supers (if any). Close up the frames in the original brood box and add one frame of foundation to replace that taken away with the queen. Remove this box to a new site at least 15 feet (5m) away on a new floor and with a new roof. All the flying bees will return to the old site and this hive will have a strong foraging force and no brood to feed. The queen will have ample room in which to lay and there should be no problems about further swarming that year.

Go back to the old colony on the new site. Remove and destroy all queen cells except one well-shaped cell which should be fairly advanced but unsealed. This portion has lost all its foragers and also its food reserves in the supers so the food situation needs watching. Feed if in doubt. This method again is a way of making increase, but if it is carried out for that purpose when no swarming preparations have started, too often the result is two very much weakened colonies. I think it best done when the bees have already started swarming preparations and when queen cells in a fairly advanced stage are seen.

Routine Examinations for Swarming

It is reasonable, I think, to examine colonies at seven-day intervals during the possible swarming period and to have a plan to deal with swarming if signs are seen. The following is a simple routine which has been well-tried and found to be effective. It is economical in extra equipment and really only needs a nucleus hive—a piece of apparatus which every beekeeper should have. The steps are as follows:

1. From May to July examine colonies every seven days looking carefully for the beginnings of queen cells. If none is found close up quietly. If you are running on double brood-chambers this can be a quick operation. Free the top chamber with the hive tool and, with one side resting on the lower brood-box, swing it up so that you have a clear view of the bottom bars of the frames in the top box. Swarm cells are almost invariably built along the bottom bars of the top box or up the sides.
2. If you see unsealed queen cells, find the queen and place her, on the frame on which she is found, in a nucleus box. Add a second comb of bees and sealed brood and a third comb of stores.
3. Shake in the bees from another comb and close the frames to normal spacing.
4. Provide the nucleus with a small entrance and move it to at least ten feet (4m) away. Do not feed for three days, but after this a little feeding will help the queen get back into lay.
5. Go back to the parent hive and examine every frame in the brood box carefully. Destroy all queen cells nearing completion. Leave other queen cells. Close the frames to normal spacing being very careful not to damage the queen cells left. Add a division board and close up.
6. Seven days later again examine all frames and select one good-sized and well-shaped queen cell—on the face of a comb if possible. Destroy all others and make quite sure that you have done so.
7. Close up.

A virgin will emerge and mate and further swarms are unlikely from the hive that year. If all goes well you will have requeened your colony with a young laying queen of your own strain and at the same time kept the old queen in a nucleus in case something goes wrong with the young queen. If you want increase, the nucleus can be transferred to a normal brood box and gradually built up to a full-sized colony. If increase is not wanted, move the nucleus gradually a couple of feet at a time until it is near the parent colony. Place a sloping board from the ground to the entrance of the parent colony. Remove the division board and shake the bees from two or three frames onto the board, being very careful not to shake the queen down. Immediately shake the bees from the nucleus down onto the board having first found and destroyed the old queen. The bees will run in happily together. Place the frames from the nucleus in the old brood-box thus completing it. If this operation is done on a fine day when bees are flying freely no trouble in uniting will be experienced.

There is a second and very easy method of uniting which can be used, not only after requeening operations but on any occasion when it is desired to unite two colonies. One evening when flying has ceased, remove the roof from the parent colony and any covering you may have over the feed hole. Place a sheet of newspaper over the cover board or quilt and make two or three pin-holes in it. Cover this with a queen excluder then an empty box into which you put the frames from the nucleus, again after having removed the old queen. If you have been unable to find her, you will find her on the excluder the next day, but it is much better to find and destroy her first. The bees will unite through the newspaper which they will chew up and the empty box and the frames from the nucleus can be removed in a couple of days.

In most accounts of swarm control methods the advice is given to examine colonies every seven days in order to be able to take action to prevent the issue of a swarm. Care should be exercised over this timing however. The seven-day examination is based on the fact that when queen cells are raised under normal circumstances the egg is laid in a specially-made queen cell and nine days later the fully grown larva is sealed in her cell where she pupates. The beekeeper is thus in a position to act two days before the queen cells are sealed if he looks at his colony every seven days. This is well and good the first time queen cells are found, but if these are all destroyed the bees will raise new queen cells from eggs one, two or three days old and even from young larvae. Suppose a young larva, just hatched, is taken (i.e. four days old from laying), the resulting queen cell may be sealed on the fifth day from the extension of the worker cell into a queen cell, the day the colony was last looked at. This may be too late.

10

Queen Rearing

Most systems of swarm control and even of management almost invariably include the phrase 're-queen with a young queen' or words to that effect, but with little or no hint as to where the young queen is to be obtained. There are occasions too when things have gone a little awry. A colony is found to be queenless or become hopelessly ill-tempered or other defects of the queen may be noticed. To the large commercial man this will be no problem for he will certainly have a queen-rearing programme as part of his management schedule. Too often the small beekeeper has no such facility and is driven to buying queens from a queen-breeder or to begging a queen cell from a friend who has a colony in the throes of swarming.

I would suggest that these are not the best ways of obtaining fresh queens. I do not disparage the qualities of professionally raised queens. Many come from good lines and are raised under optimum conditions. They may give good results. Others may not be so good and a few are poor. They may have travelled through the post and have suffered the tender mercies of the postal system which can give rise to introduction difficulties, and the first cross of daughters from line-bred queens can have undesirable characteristics. Some come from abroad, although I am happy to note that there are increasingly stringent legal restrictions on the importation of live bees. These are imposed because of the very real danger of importing disease with worker bees.

Check tests for the year 1978 carried out by Ministry scientists showed that of the bees examined 47.7 per cent were infected with Nosema disease and 52.3 per cent were found to be clean (M.A.F.F. Bee Health Statistics 1978). Practically all the infected bees came from the USA and Italy.

Obtaining a New Queen

The idea of raising a few queens each year seems too difficult and too mysterious and many beekeepers shy away from the task. It is true that the production of queens on a large scale requires the employment of considerable capital, special equipment, very tight timing and backing by an efficient business organization. However it is perfectly possible—and not at all mysterious—for the beekeeper with one or two colonies to raise his own queens. It is even possible to raise a replacement queen where only one colony is maintained. Still, the more colonies that are available to choose from, the greater will be the possibility of raising good queens. There is a further point to be borne in mind with regard to bought queens. The beekeeper in Loamshire who buys a queen from a reputable queen-breeder in Mudshire at the opposite end of the country can easily be disappointed. He may find that the new queen does not prosper in his local climate. If queens are raised from colonies of good characteristics in local conditions satisfaction is more likely.

Perhaps the simplest way of all of obtaining new queens is to wait until a colony is building queen cells and then to deal with them. The queen can be killed and a well-shaped queen cell left to hatch and mate, other cells can be distributed to colonies needing re-queening. The colony can be artificially swarmed or the colony can be sacrificed and divided into a number of nuclei each with a queen cell. Some of these actions can be acceptable as *ad hoc* methods of dealing with a colony about to swarm, but they are far from ideal as queen-raising exercises. The beekeeper has little control over what happens and when, and they should really be considered panic measures.

The next simple way is to remove the queen from a selected colony, either in a nucleus or by confining her in a part of the hive away from the brood nest. The bees will do the rest. They will provide you with queen cells to be used as you wish. It is possible to exercise some choice in this method by selecting a comb with eggs and young brood and cutting away the comb below a patch of eggs or very young larvae. If this is done some hours after the colony has been made queenless the bees will often draw out some of the cells on the bottom edge of the comb into queen cells.

The Miller method is to use two colonies—one to supply the eggs and larvae to be your future queens (this will be your best colony) and another to raise the queen cells.

An empty brood frame is prepared by fixing to the top bar either four triangular pieces of unwired foundation with the points reaching about half-way down the frame or a single strip of foundation cut in a zig-zag with four triangular points. Place this prepared frame in the brood-nest of the good-quality colony. In about seven days the bees will have drawn out the strip(s) of foundation and the resultant comb will contain brood and eggs with the eggs and young brood towards the edges. Go to your cell building colony. This should be populous and, if possible, have an abundance of young bees. Remove the queen to a nucleus box with a frame of young brood and frames of food.

Leave the colony for three hours, then take your prepared frame from the breeder colony and *gently* brush the adhering bees off with a feather. Trim the edges with a warm knife or warm scissors so that the youngest larvae are at the edges. Be extremely careful with this comb. It is unwired and very fragile. Place the comb in the space in the cell-building colony where you have removed the frame with the queen and close the frames together. In ten days time there should be a supply of queen cells for you to distribute to nuclei or to colonies that have been de-queened the previous day. Cut the cells out with a generous piece of surrounding comb so that you have something to hold other than the queen cell itself. Stick the queen cells to the face of a comb, preferably one which has some brood, by means of the piece of comb.

Queen-rearing for Small-scale Beekeeping

One well-tried method of getting a few queen cells raised without removing the queen from the colony is to encourage a strong colony early in the season with plenty of young bees. This is a good method for the small-scale beekeeper and if your efforts fail you are still left with your colony unimpaired.

The method involves working on a double brood-chamber. You should aim to have a large and prosperous colony towards the end of May. If you have wintered the bees on one brood-chamber, a second box should have been added as early in the year as weather conditions permitted. Add a super or, better still, two supers before the end of May. Now sort out the frames in the brood-boxes—all the unsealed brood in one box with a good frame of pollen and with any space filled with sealed brood. The pollen frame should be in the centre and any sealed brood towards the outside. Put all the other frames in the remaining brood-chamber together with the queen. It helps if you, put the queen in a matchbox with a couple of workers when you find her during sorting. Keep the matchbox in a warm pocket so that you are quite sure you know where she is when it comes to putting her back.

Re-assemble the hive as follows: brood-chamber with queen on the floorboard; excluder; supers; another excluder; and finally the brood-chamber with the young brood and the pollen comb. Complete with cover board and roof as usual. The young nurse bees will come up to the top box to look after the unsealed brood. They will find themselves separated from the queen by two queen excluders and two supers and will get less than their normal ration of queen substance and will raise queen cells. You have placed a full pollen comb up near the eggs and young larvae so the nurse bees will have a handy supply of protein in order to keep their brood food glands well supplied.

As the queen cells raised will be in the top box they will be conveniently placed to be dealt with without dismantling the whole hive. If all goes well, you will find a small number of queen cells which can be distributed to mating nuclei or the brood-chamber can be split into two or three nuclei, each with a good queen cell. Do not be too greedy and make too many sparsely populated nuclei. It is better to destroy a few queen cells and retain two or at the most three nuclei with plenty of bees in each. These can be retained as mating nuclei through the season if you wish to go on raising queens, united to colonies needing re-queening after the young queens have mated, or they can be built up to full-sized colonies.

Grafting into Artificial Cell Cups

In 1888 G. M. Doolittle published his method of queen-rearing. It consists basically in transferring young larvae from worker cells into specially made queen cups. These are then given to de-queened colonies to be drawn out and the larvae fed. The system is called the 'grafting' method. It is eminently suitable for large-scale queen-rearing and is used, with variations and permu-tations, by most of the big producers. Whatever alterations are made I think it is fair to say that they all owe their origins to Doolittle's original conception. However it is a method which can quite easily be used by the small-scale man too and is well worth trying. The following is a very brief account of a method which might be suitable for the beekeeper wishing to raise not more than, say, fifty queens during the season.

Cell Cups
When only a few cups are needed the 'play' cups often found at the bottom edges of brood combs have been used, but it is much better to make one's own cups from clean beeswax. A forming stick is needed for this. A piece of hardwood dowel rod about $\frac{3}{8}$ inch (9mm) in diameter serves well. The last inch should be tapered to about $\frac{1}{4}$ inch (6mm) and the end rounded off. Make the forming stick as smooth as possible with fine glasspaper. The purpose of this and of the taper is to facilitate removal of the cups. A small glass test tube of about the same dimensions can be used. Although not tapered, the cups seem to come away from the glass quite well. The forming stick is soaked in clean cold water for a few minutes, the surplus water shaken off and the end is then dipped into molten (just molten) wax to a depth of $\frac{3}{8}$ inch (9mm). Withdraw quickly and repeat four or five times, then plunge the former into cold water until the wax has solidified. The cup should then come off with a gentle twist. Repeat until you have enough cups. A few extra is a good idea in

case you crush any in handling. Large-scale beekeepers have ten or more formers mounted on wooden strips and even ten such strips fastened together so that a hundred cups can be produced at a time.

The cups have to be fixed to a bar of wood of such a length that it will fit nicely between the side bars of one of the frames you normally use and rest on small wooden blocks nailed to the side bars (or slots can be cut into the side bars into which your bar of cups will rest). The cups should be spaced so that the resulting cells can be cut out individually without interfering with their neighbour. About $\frac{7}{8}$ inch (22mm) is ample from centre to centre. One frame can take up to three bars of cells which should be fixed so that there is at least $1\frac{1}{2}$ inches (38mm) between the bottom of a bar and the top of the bar below (or the top of the bottom bars of the frame).

Now to fix the cups to the bar with molten beeswax. Acquire

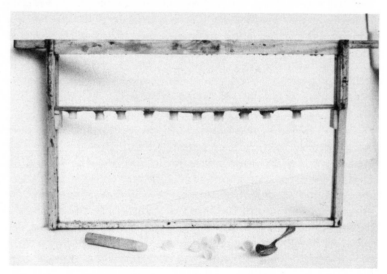

Figure 29.   Frame of bee cups.

an old teaspoon past its better days. The boss of the house usually has two or three in that drawer in the kitchen and you ought to be able to obtain one either by stealth or cajolery. Bend the sides of the spoon up to make two lips to make a convenient pouring tool. Scoop up a spoonful of molten wax and run a streak of it down the centre of the bar. Before it sets gently, set your cups into it at the required distances. After the wax has solidified a little more melted wax can be run at the sides of and between the cups. This will give added strength. Some people like to fasten the cups to small squares of wood which are, in turn, stuck to the bars, on the grounds that the wooden squares give something firm to hold. It is certainly true that great care must be exercised in handling the queen cells. The virgins inside will be at the prepupal or pupal stage and will be soft and delicate.

Each cup must now be provided with a larva from a worker cell from your breeding colony. The age of the larva is most important. They should be not more than twelve hours old. At this stage they are only a little longer than the egg from which they have hatched and shaped like an open letter 'C'. If the head and tail ends have met to make a circle they are too old. The tool for transferring the larvae can be a 'grafting needle' made for the purpose and sold by the bee appliance manufacturers or a small quill cut to a point and shaved down (a toothpick?) or even a match shaved down, pointed and the tip bent over between the teeth. On the whole I think a proper grafting tool is best.

Many experienced beekeepers transfer the larvae to dry cups, but it is easier to prime each cup with a little royal jelly diluted with water first. If this is done the tiny larvae float on the drop of liquid at the base of the cup quite easily. Lower the grafting tool down a side wall of a cell and slide the tip under the larva at the back, i.e. under the closed part of the 'C'. Lift the tool and the larva will come up resting on the tip of the tool. Lower into the pool of diluted royal jelly in a cup and withdraw the grafting tool gently, sideways and up. The larva will be left floating in the queen cup.

The whole process of grafting should be carried out in a warm atmosphere—a shed or, in the case of an out-apiary, in a car with the windows closed. Ideally the conditions should be humid. This is to guard against chilling and dehydration.

Cell Building

The bar with artificial cups, each with its own larva should now be given to a cell-building colony. This must satisfy three criteria. It must be queenless, have plenty of young bees and have ample pollen stores close to where you propose to insert your frame of queen cups.

It must be emphasized that the above is no more than the basic principle of producing queen cells by the grafting method. There are many elaborations of the system which the reader will find in the books quoted in the *Further Reading* section. This is not the place to go into them but I would strongly advise that, after reading a selection of the references, a trial of the simple method be made. It is unlikely that all the queen cups will be accepted at the first go. Try it out and stick to one simple method until expertise and success is achieved. After this, more complicated and sophisticated methods can be tried. However do not try a mish mash of several methods. Keep it simple.

Mating the Virgin Queens

Giving ripe queen cells to queenless colonies is not the surest or the best way to re-queen. Quite often a colony will only accept what they have been used to. A colony with a laying queen, if de-queened, may only take kindly to another mated queen and may pull down a cell, and the bees' labour, and yours, will have been in vain. A far better plan is to get your virgins mated from small mating nuclei. You then have the opportunity to see whether the virgins have successfully mated by observing eggs in the mating nucleus (or the absence of them!). Mating nuclei can be quite small if that is what is wanted. Many professionals use 'baby' nuclei consisting of two frames of comb about 5 inches (13cm) square plus a frame feeder of the same size. This is a convenient arrangement when queens are to be produced on a large scale for sale. As the queens are 'proved' (i.e. seen to have been properly mated) and sent off by post to buyers, the nuclei become free for new queen cells. A more practical way for the more modest beekeeper who is rearing a small number of queens is to make up nuclei consisting of one standard frame of sealed brood with adhering bees, one frame of food and one empty comb (with no eggs or young larvae). If it is possible, the mating nuclei should be sited more than two miles from the home apiary. If this is not possible, shake in the bees from another comb, otherwise the older bees will fly back to their parent apiary and leave your nucleus a little denuded of bees.

Two days after making up the nucleus fasten a ripe queen cell by its base to the frame of sealed brood and leave undisturbed for fourteen days. A useful variation is to divide a standard brood-chamber into two equal compartments by a sliding board fitted into a groove cut into the side walls. This should fit closely to the floor too so that there is no access between the two halves. It is probably safer to screw a flat floor to the bottom of the box and to make small entrances on opposite sides. Each half should have its own top cover and in this case I think canvas quilts are better than cover boards. A single standard roof covers the lot.

If built up to five frames, nuclei can be wintered safely and are available in spring for any colonies which have become queenless during the winter. The possession of a young mated queen established in a nucleus gives a very comfortable feeling of having something to fall back on in an emergency. I cannot help feeling that this is the answer to devotees of buying foreign bred queens who say, quite rightly, that queens cannot be raised early enough in the year in our cool temperate climate to meet such emergencies.

Two methods are recommended for uniting nuclei with queenless or de-queened colonies. The first, more suitable for use during the active season, and even more so if there is a nectar-flow, is to sprinkle the bees in both nucleus and colony with flour or spray them with sugar syrup and then simply make space in the colony and insert the frames from the nucleus directly. In order to avoid loss of flying bees, the nucleus should have been moved gradually, not more than four feet at a time, closer to the colony until they are adjacent. The flour or syrup will mask the natural scent of the bees and they will be so busy cleaning themselves up that they will unite peacefully.

The second method is even simpler. Having got the two lots close together, open the queenless lot quietly in the evening and cover the brood-chamber with a sheet of newspaper. Prick a few holes in this with your hive tool and set another brood box on it into which you place the frames from the nucleus. In 24 hours the bees will have chewed through the paper and will have united. Both methods leave you with hives not completely filled with frames and you will have to sort this out at the first opportunity. It may give you the chance to get rid of those mucky frames at the outsides of the brood-chambers that you have meant to replace anyway.

11

The Honey Harvest

There are often occasions quite early in the year when a spell of good weather will enable bees to fill a couple of supers with capped honey. It is natural to want to remove your first honey and it is a good plan to do so. The honey from nectar gathered in the spring months will be from different sources and will have different flavour and characteristics from that gathered in high summer. If you do take early honey, keep a close eye on the food situation of the colony. A cold spell in early summer can put a colony in jeopardy. You should extract from the full combs and return them to the bees, wet, late in the evening. Do this quietly as late as possible.

Shake and Brush Method

The easiest way to deal with a few combs of spring honey is the 'shake and brush' method. Smoke the colony as usual. Remove the roof and invert it at the side and a bit to the rear of the hive. Place an empty super in this with a cover cloth over it. Give a few extra puffs as you take off the inner cover. This time it will not matter if a little smoke goes down between the frames. If the combs appear to be well filled, lift out the first comb and brush

(a)

Figure 30.   (a) Brushing bees from a comb. (b) Top, the old traditional tool, a piece of goose wing and, below, a modern bee brush.

(b)

off the bees with a soft brush with long hairs or a piece of poultry wing. A goose wing is the traditional tool. Put the comb, cleared of bees, in the spare super and cover it with your cover cloth. Lift the next comb halfway out and in the space you have made by removing one comb give it a sharp downwards jerk stopping abruptly at the bottom of the jerk. This will dislodge most of the bees and the rest can be brushed off as before. Repeat through the supers until you have all the combs cleared. This need not be the complete contents of a super. Just take full combs of sealed honey. You will be returning the extracted combs in a few hours to fill the spaces.

This shake and brush method is suitable early in the year and is made easy when there is a little nectar coming in on a fine warm afternoon and the bees are flying freely. It is altogether another matter at the end of the season when the nectar-flow has terminated—suddenly as sometimes happens. At this time the bees, not surprisingly, resent being robbed, and the shake and brush method should only be used by the resolute and well protected and is not recommended for the beginner. The response of the bees can be quite startling.

Special mention must be made of the honey from rape. All brassica types of honey crystallize rapidly and rape honey particularly so. It will crystallize in the comb and the safe thing to do is to extract combs you know to be of rape honey as soon as they are capped. Some people like to extract as soon as the combs are three-quarters capped and do not wait until complete capping. Combs of rape are not good winter stores for the bees either. The set is usually too hard for them to make use of it. Do not leave it too long in the settling tank either. Many a good beekeeper has come downstairs in the morning to find a half hundredweight of solid honey in his settling tank. The flavour and aroma of rape honey, to my mind, is not pleasant but the set is of fine grain and is useful for blending with strong-flavoured honey. Like any other crystallized honey it can be softened, even liquified, with the application of gentle warmth.

The Main Crop

At the end of the season it is preferable to clear supers of bees in such a way to minimize disturbance. It is important not to stir them up too much at this time of the year as they can easily be provoked into a robbing session and this is best avoided. Robbing is difficult to stop and the colony robbed (usually the weakest) is often robbed right out. They will fight of course and losses of bees on both sides are sad to see. There are three principal methods of clearing bees today: chemical repellents; forcibly blowing the bees off; and the use of clearer boards.

Chemical Repellents

The notion behind using chemical repellents to clear bees from supers is to use some kind of volatile substance with a powerful odour which the bees dislike and will move away from, but which will not harm them. The method is a time-saver where bees are kept in an out-apiary because only one visit is needed. The supers can be cleared and taken away in one session.

The method of application is to make a shallow cover the size of your normal cover boards. This needs to be about one inch (25mm) deep instead of the usual $\frac{1}{4}$-inch (6mm) bee-way. The top can be made of soft insulation board, hardboard, or any other suitable light material to hand. If the material used is non-porous, tack a cloth tightly to the inside, covering it completely. A teaspoonful of the repellent diluted with water in the proportion of about one part repellent to three parts water is sprinkled evenly over the cloth or the insulation board. If the mixture is too strong it seems to have the reverse of the desired effect. Instead of moving away the bees hang motionless on the combs.

Smoke the bees well and when the inner cover is off, smoke again to get the bees moving down then cover with your repellent impregnated board. Leave for a few minutes. If the bees have moved down take the super off and place the board on the next super down and so on. If you have two or more boards you can move down a row of hives coming back to the first and down the line again taking off supers.

The method is most successful in warm, slightly humid weather. The bees are reluctant to go down in cold weather or if there is any large amount of unsealed honey. Effects vary with the strain of bee and weather conditions and the downwards drift of the bees may be slow. They do not seem to be upset in any way by the repellent and merely quietly move away from it.

It has been claimed that the use of chemical repellents can taint the honey but I do not think this need be so if application is made carefully and recommended quantities are not exceeded.

The three chemicals mostly used are:

1. *Carbolic.* Used in the past, but no longer in favour because of its pungency, which is very clinging. It remains on the wood of the frames for some time and will taint the honey.

2. *Proprionic anhydride.* Expensive and sometimes difficult to obtain. It is extremely volatile and seems to 'evaporate through the cork' (like the best whisky but for different reasons!). At one time there was a scare that it is carcinogenic, although this has not been proven to be true. It is recommended that some kind of bellows arrangement is attached to the repellent board so that the fumes can be blown down between the frames. This would seem to be an additionally cumbersome piece of apparatus.

3. *Benzaldehyde.* This is artificial oil of almonds. Stocked by most of the appliance manufacturers if unobtainable from the local chemists who tend to produce dirty looks if asked for anything other than proprietary headache tablets. This is the most widely used and successful of chemical repellents. After using it the bees do not get cross as sometimes can be the case with proprionic anhydride.

Forced Air Blowers

This is a method much used by the large-scale beekeepers, especially in the United States. It has the advantage of being a one-visit method of clearing supers and can be used in most weather conditions. The capital outlay would hardly be justified for the beekeeper on a small or a medium scale. The apparatus consists of a small petrol engine driving a fan. A compressor is unsuitable. What is wanted is a large volume of air moving fairly rapidly, but not a blast of compressed air issuing from a small nozzle. Supers are taken off the hive and the bees literally blown off the combs. The effect must be a bit shattering to the bees, but it seems that they find their way back to the hive when they are less disorganized. I imagine that a similar effect could be obtained by using the reverse end of a domestic vacuum cleaner if the hives were in cable distance from an electric point. I have had no personal experience of this way of clearing supers so, in all honesty, cannot recommend nor decry the system.

Clearer Boards

Clearer boards are usually inner covers with one, or sometimes two, holes into which an 'escape' is inserted. The escape is a device through which bees can pass one way but not the other. If such a clearer board is inserted below a super, the bees will pass through the escape on their way down to the brood chamber and will be unable to return. The clearer board can be put on with little fuss and the cleared super removed later with an equal lack of fuss.

The older, and still the favourite, escape is the 'Porter'. It is made of thin tinplate and takes to pieces for cleaning. The top plate has a central hole below which are two sets of paired weak springs arranged in a 'V' but not quite touching. Bees can push through, overcoming the weak pressure of the springs. The underneath part is shaped into a shallow trough longer than it is wide and slides in grooves in the top piece, and carries the springs. Nowadays they can be bought made of plastic material and there are other plastic devices giving more than two exits, but based on the spring principle.

Another popular clearing device is the 'Canadian' board. Two alternative methods of construction are possible. They are made to the exact measurements of the top of your hive and with edging strips on both sides of $\frac{5}{16}$ inch (8mm) wood, or a little more to give beespace top and bottom. If the cover is made of boards at least $\frac{5}{16}$ inch (8mm) thick, a gap of an inch (25mm) or so is left between the boards in the middle. On the top side two pieces of sheet metal are nailed so that they fit tightly against the edging strips, but stop short of each other by about an inch (25mm). On the underneath side a piece of wire mesh or perforated zinc is fastened so that it covers the whole of the gap between the boards, except that this time it does not quite reach the edging strips but stops short of it by about $\frac{5}{16}$ inch (8mm) at each end. The tunnel thus formed will be at least a bee-space

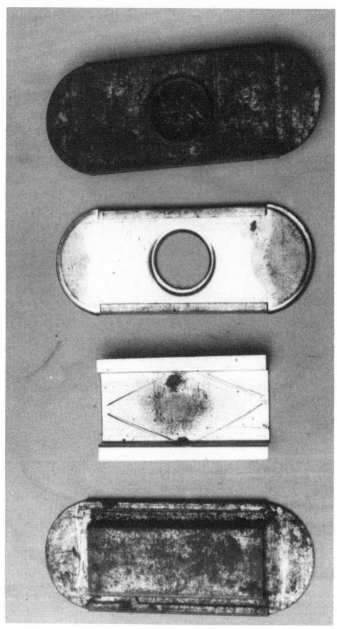

Figure 31. Porter bee escapes. Far left and far right are top and bottom pieces respectively. The two centre pieces show the escape taken apart to reveal the springs.

(a)

(b)

Figure 32. Top (a) and underneath (b) views of a Canadian clearer board.

deep and will run from the central gap between the metal plates almost to the edging strips on two opposite sides.

If a clearer board is made of ply or other material less than $\frac{1}{4}$ inch (6mm) thick, i.e. less than a bee-space, cut a central hole similar to the usual Porter escape hole and on the underneath side fasten two wooden strips $\frac{5}{16}$ inch (8mm) square running across the board on either side of the hole and tack a strip of wire gauze or perforated zinc over them to just short of the edging strips.

In either method of construction the bees will come through the central gap or hole and into the brood-chamber via the gaps at the end of the tunnels, and do not seem to have the sense to go back the reverse way. I think it most likely that they would learn to do so if the clearer were left on for any length of time, but clearing is usually complete in 24 to 48 hours and the clearer board is then removed.

There does not seem to be any clear difference in efficiency between the Porter or the Canadian type of clearing, but the Canadian has no moving parts to be propolized and neither does it need cleaning. On the other hand a Porter escape will be propolized up fairly quickly and will not work well unless it is kept clean and slightly greased.

The Canadian system seems to work equally well if, instead of a central hole, holes about $\frac{1}{2}$ inch (13mm) in diameter are cut in diametrically opposite corners of the board and the underneath of the holes are covered by small squares of perforated zinc of about two inches (50mm) held clear of the board by two strips of $\frac{5}{16}$ inch (8mm) wood, arranged to leave a gap for exit purposes.

It is more efficient to keep one or two clearer boards specifically for the purpose of clearing instead of cannibalizing an inner cover and perhaps breaking the propolis seals the bees have made.

Any tiny holes or cracks in the supers or between them and their neighbours will be quickly found by the bees immediately you start clearing and if there are any, you may be disappointed to find that your crop has been spirited away overnight. Make quite sure everything is bee-tight. A lump of 'Blue Tack' in your equipment box is a useful hole stopper. It is just sticky enough to remain in a hole temporarily until you are able to make a more permanent repair.

12

Extracting the Honey

I t is not possible to over-emphasize the importance of cleanliness at every stage of extracting, storing and bottling honey.
If the honey is intended for domestic use, the beekeeper will take
pride in bringing a pure food to the table, handled and bottled in
hygienic conditions. If some of it is intended for sale, then such
conditions are legal requirements (see *Appendix 3*).

The Extracting Room

The room in which honey is extracted should be clean and free
from dust and should have a supply of running hot and cold
water. For many beekeepers the kitchen becomes the 'extracting
room' once a year and as I have yet to meet the housewife who
would tolerate dirt and flies in her kitchen the requirements of
cleanliness are no problem. Bees and wasps are strongly
attracted by the smell of honey being extracted, so the room
should be bee-tight. When small numbers of combs are to be
dealt with it is possible to take the cleared combs off the hive
during the day, extract at night and return the wet combs to the
bees next day and so avoid any commotion.

However when dealing with the main crop this is not possible and the supers awaiting extraction will have to be stored in clean bee-tight conditions. Use plenty of newspaper on the floor to catch the odd drips of honey. This will help to keep the working space free of honey and will facilitate mopping up. Wear clean overalls when working and when the time comes to mop up, do so thoroughly. Nothing will mar domestic bliss more quickly than sticky door handles.

All equipment used should be thoroughly washed in hot water and dried before and after use.

Honey leaves the combs most readily if it is warm. Straight from the hive is best, but this is not always possible. A stack of supers over an empty box in which there is a low-wattage light bulb (not more than 40 watts) will keep sufficiently warm. Take frames out of the lowest super and find room for them up above so that there are no combs immediately above the light bulb.

Uncapping

Before the honey can be extracted the wax cappings over the

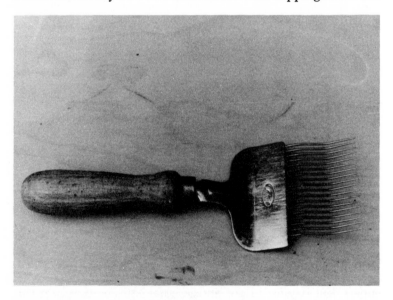

Figure 33.   An uncapping fork. The sharp tines are inserted and run along just below the cappings of the comb.

cells must be removed. For small and medium quantities I like to use an uncapping fork. The points of the long tines are run along just below the cappings which come off in strips.

For a greater number of combs a knife or an uncapping 'plane' is probably quicker. Both can be bought with electrically heated blades, but a stainless steel domestic knife with a thin blade with scalloped ground edges does just as well. The uncapping plane is rather like a large safety razor and works in the same way. The blades are adjustable for depth of cut.

Frames to be uncapped are held upright resting on one lug and inclined towards the uncapping instrument. The cappings then swing clear of the cut surface and fall into whatever cappings container you are using. If you use metal ends these must be removed from the frames before uncapping and loading into the extractor.

Dealing with the Cappings

A receptacle must be provided to catch the cappings as they fall from the combs. The simplest is a large meat dish, clean of course, with a piece of wood across it on which the frame lugs can be rested. From time to time accumulated cappings and honey can be tipped onto a piece of nylon, muslin or bolting cloth stretched over a large bowl. This will do for a few frames, but something more efficient is needed for larger quantities. One slightly more sophisticated method is to use what is called a 'Pratley' uncapping tray. This is a double tray about 22 by 14 inches (56 by 36cm). The top section slopes towards one end which has an outlet covered by a perforated zinc baffle, and has a cross bar upon which to rest the frame lugs when uncapping. The lower portion of the tray forms a water jacket and has a small immersion heater. Cappings fall on the upper tray and are melted by the warmth from below and the mixture of honey and molten wax flows through the outlet and into whatever receptacle is placed below it. The wax floats and in the morning will have solidified into a cake below which is the honey.

Yet a third method is to make a box to the dimensions of your supers, but having a floor and internal walls so that bees can have free access to the inside, but the box itself is a bee-tight fit onto the hive. Fit either a cross-piece or recesses in the inner

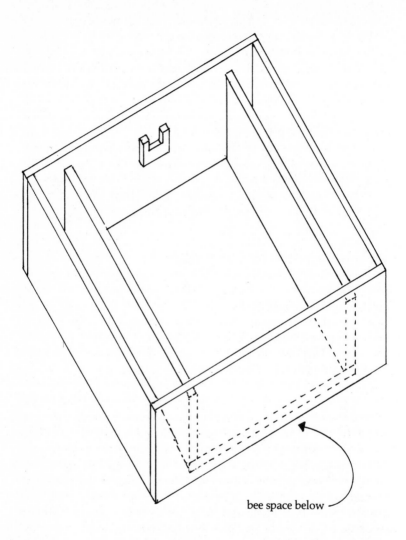

bee space below

Figure 34.   Uncapping tray made to the size of the hive. In use frame lugs are rested in the U-shaped cleat. Cappings fall onto the solid floor. When placed on a hive and covered with an ordinary crown board, the bees can come up the side gaps and will clean any honey.

walls on which to rest the frame lugs. Uncap in the usual way allowing the cappings to fall into the inner box. After you have finished uncapping, in the evening quietly take off the roof and inner cover from a hive, place your cappings box on as you would a super and close with your inner cover and roof. I have used such a box for many years and have had no trouble with excitement or robbing. I take care to see that there are no smears of honey on the outside of the box so that there is nothing to attract inquisitive bees from other hives. The bees will clear up every drop of honey and leave you with dry wax to render.

Extracting

The uncapped combs are loaded into an extractor. Basically this consists of an outer barrel made of metal or rigid plastic, inside which is a pivoted cage which holds the frames. The cage can be rotated by hand with a simple train of gears or by a power drive. As the cage rotates honey is flung out of the combs by centrifugal force and runs down to the base of the barrel from which it can be drawn off via a honey tap or valve. This is sometimes called a honey 'gate'.

Two types are available. In the tangential type the frames are held at a tangent to the circumference of the barrel, i.e. at right angles to the radius. Each face of the cage is fitted with wire mesh, nowadays coated with white plastic for ease of cleaning. Various sizes are available, taking up to four deep or eight shallow British Standard frames. When the cage is revolving at speed a good deal of 'g' is generated, so it is important to load combs of comparable weight diametrically opposite each other. If this is not done an uncontrollable wobble will develop. The weight of honey on the inwards facing sides of the combs exerts considerable pressure on the midrib when they are spun and damage could result if the cage is rotated too fast. It is advisable to rotate fairly slowly until the bulk of the honey in the outer sides is extracted. This happens quite quickly. Then reverse the combs. Rotate again, slowly at first and then more quickly to get complete extraction. Reverse the combs once more and complete extraction of the first side.

It is usual to load the frames into a tangential extractor with the bottom bars first in the direction of travel on the grounds that extracting is easier because the cells slope upwards slightly from the midrib. In practice the difference, if any, is so small as to be immaterial. Indeed, there is mathematical argument to the effect that the centrifugal force generated acts parallel to the slope of the cells whether they are facing forwards or backwards to the direction of revolution.

The radial type of extractor has an internal cage in which the frames are loaded radially from the central spindle. In this case it is important that the top bars of the frames are on the outside

Figure 35. Tangential extractor: interior view showing handle and cage in which the frames of honeycomb are placed.

and the bottom bars towards the spindle. The method is not so efficient as the tangential type, i.e. the rate of honey extraction is slower, but on the other hand both sides of the combs are extracted simultaneously so that no reversing is necessary and the capacity, in terms of frames, is usually greater. The system lends itself readily to motorization.

Space is provided at the bottom of all extractors for the accumulation of honey which has to be drawn off from time to time. It is convenient to have a strong wooden stand not less than twelve inches (30cm) high. The extractor can rest on this and even be screwed down when lugs are provided on it. A standard 28 lb lever lid tin can then be set down under the tap for filling.

After extraction the wet combs are replaced in their supers and you have the choice of either giving them back to the bees to

Figure 36.   Wooden trestle to hold 28 lb tin.

clean up, which they will do with alacrity, or storing them wet for use next year. It is surprising how much honey is left behind in a well-extracted comb and it seems a good idea to let the bees have this.

The area of the bases of the cells plus the area of the side walls in a super of nine wide spaced British Standard shallow frames is about 100 square feet (9.5 square metres) and no amount of prolonged extracting will get this area dry because of surface tension. In addition, each cell will be left with a thin hemi-spherical bubble of honey also held there by surface tension. A trial run with the combs from a number of reasonably similar supers gave the following figures which are averages for one super of nine combs.

Table 7.   Honey Extraction Figures from Nine Combs

Dry Weight	Full Weight	Honey Extracted	Weight after Extraction	Honey Left in Combs after Extraction
9.7 lbs	37.6 lbs	25 lbs	12.5 lbs	2.9 lbs

If the combs were all drone combs, the wettable area would be about 80 per cent of what it is with worker comb and the surface tension factor of the bubbles at the cell mouths would be only about half that of worker cells. On these figures one would expect the retained honey in a super of nine combs to be about 2 lbs (0.9kg) compared with 3 lbs (1.4kg) for worker combs— but still a significant figure.

Straining

During the process of extraction, small pieces of wax, cappings and other debris will have become mixed in the honey and air bubbles will have been formed by the whirling motion of the extractor. These have to be removed by straining and allowing the honey to stand in a tank so that the air bubbles can rise to the surface. A settling tank, sometimes called (erroneously) a 'ripener' is a tall cylindrical tank usually about three times as high as it is in diameter, made in either tin plate or heavy duty polythene. They are fitted with an upper straining unit incorporating a metal mesh to retain the wax pieces. A piece of straining cloth is tied over the mouth of the tank and allowed to hang down inside in a loose bag—but tied securely at the top!

A mesh of 54 to the inch is satisfactory and nylon strains much faster than cotton. Honey from the extractor is tipped into the top strainer section and after it has passed through this and the nylon cloth below it, it is allowed to stand in the tank for 24 to 48 hours. After this time the air bubbles will have risen to the top and the cleared honey may be drawn off via the honey tap at the bottom into 28 lb lacquered tins. It sometimes happens that

*Figure 37.*Extractor (right), small settling tank (left) and cone-shaped perforated zinc strainer (middle).

there is incipient crystallization in the comb and if this is the case the honey will be very slow to pass through the straining cloth. If this happens it is better to draw off the honey from the extractor straight into 28 lb tins.

At a convenient time later scoop the bits of wax off the top and warm the tins to about 95°F (35°C) for a short time. The honey will then pass through the cloth. After 48 hours again skim the top. This time it will merely be froth to take off. The honey is then in good condition for storing which should be in a cool, dry place, ideally at a steady 50°F (10°C).

Storage

While in storage the honey will crystallize. It is best to let it do so in the tins and then deal with it as described in the next section. It *can* be run off into bottles straight from the settling tank, but initial crystallization is often uneven and when set hard the honey will shrink away from the sides of the bottle and look unsightly. This is known as frosting. There is nothing wrong with honey which does this. The crystals by the glass are larger than the bulk and the effect is not attractive.

Do not store in tins where the lacquering inside is faulty. Honey should never be allowed to come into contact with iron. If it does it forms an unpleasant black substance, iron tannate. If you have, or acquire any 28 lb tins which look doubtful after a good wash, do not throw them away. Put a large polythene bag inside each and fill the bag with honey. The tin will give rigidity and the polythene will stop the honey coming into contact with the suspect spots in the tin. Secure the bag with one of the wire and paper clips used for closing food bags in freezers. Although it is true that good honey in proper storage conditions will keep for a long time, I think it is best bottled and used or sold within a year. Legal requirements formulated by the E.E.C. lay down minimum levels for the diastase content of honey and the maximum permitted level of HMF (hydroxymethylfurfural) and there is a possibility that the former might fall and the latter increase with prolonged storage, although I do not believe that either would affect the wholesomeness of the honey in the slightest degree.

Once the honey is packed into 28 lb tins it can be safely left in a cool place until you are ready to bottle it. When you come to do this the honey will almost certainly have set into a firm intractable mass. Gentle warmth will soften it to a pourable consistency with little of the crystallization being melted. Higher heat will melt the crystals completely and return the honey to the 'clear' condition. Warmth at about 85°F (30°C) will soften the hardest set honey in four days and 120°F (50°C) will melt it completely in two days. Even so, there will be plenty of tiny crystals in suspension and the honey will re-crystallize in the bottles.

Crystallization is always initiated round small solid particles in suspension, e.g. pollen grains or crystals too small to be seen. If you want to spoil your honey then by all means heat it strongly and filter it through extremely fine filter material. This will remove all these small natural particles and will leave you with a very attractive clear honey that will have considerable sales appeal.

Warming Box
A warming box is a very useful piece of equipment for any beekeeper. Figure 38 illustrates a basic design I have used for many years. No dimensions are given because size is immaterial except that the inside of the inner box must be large enough to take two 28 lb tins side by side on supporting slats with space underneath for a heating device. In my case this consists of two ordinary electric light sockets wired in parallel. I can then switch bulbs of differing wattages to get the degree of heat I want. Two 40 watt bulbs seem about right for most types of honey. I like to fit a sheet metal baffle over the bulbs (but not too near) to disperse the rising heat.

A small hole in the side of the box about two thirds the way up through which a long-stemmed thermometer is inserted enables me to keep a close eye on the temperature.

I think a refinement might be a small black-bar heating element coupled to a thermostat instead of the light bulbs but have been content to muddle along with my light bulbs. I can at least see immediately that they are working.

Material for the box could be wood boards, blockboard, chipboard or plywood. I find $\frac{3}{8}$ inch (9mm) ply suitable and not too heavy. Many people make their boxes with fixed tops and front opening hinged doors. I think it is more practical and there is less heat loss if the opening is at the top. If, however, the lid does not fit well, then there *will* be heat loss.

As the sketch shows the box is double skinned. Leave about three inches between the outer and inner walls and pack the space with insulating material. Screwed up sheets of newspaper

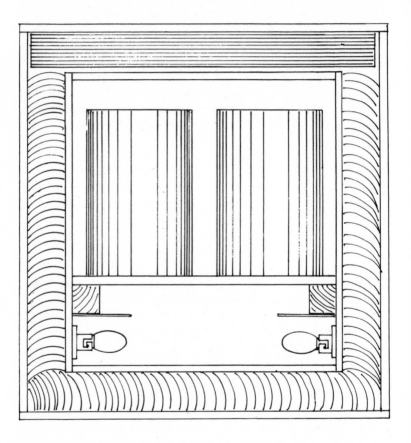

Figure 38.   Honey warming box. No dimensions are given because this can be made to any convenient size—to take two, three or only one 28 lb tin.

packed fairly tightly will do and odd or broken pieces of expanded polystyrene ceiling tiles are better. This material is often used for packing in cartons round fragile articles like radio sets. Line the inside of the lid too with felt or more expanded polystyrene.

When the honey is warmed and softened, gentle stirring with a paddle or a clean stick will make for even consistency.

The honey is now ready for bottling. The tins can be emptied into a washed and dried settling tank and the bottles filled from the gate at the bottom of the tank. Fewer air bubbles will be introduced if the bottles are held close to the gate and at an angle until they are nearly full.

Direct Bottling

Instead of transferring the honey to a settling tank, it can be drawn directly from the 28 lb tins into bottles in the following way. Drill a hole in a 28 lb tin lever lid as close to the rim as possible and into this fit a small honey gate (obtainable from the appliance manufacturers). Drill a second hole of about $\frac{1}{8}$ inch (3mm) diametrically opposite.

If the lid is now fitted to a full tin of honey and the tin is laid horizontally in a trestle which can be tilted, then pouring can be done conveniently at a table or bench. Make quite sure that the lid is a tight fit and for good measure tie twine round the lid and tin. Some 28 lb tins have overlap lids instead of the usual lever lids and in spite of standardization not all lids fit equally tightly so care is needed.

Of course, not all your honey need be stored in tins and you will certainly want to get some of it into bottles as soon as possible. This can be done after extracting as soon as the honey has cleared in the settling tank.

Use standard 1 lb glass bottles with screw lids and clean wads.

13

Diseases, Pests and Other Nuisances

Bees, like all other animals, are sometimes subject to diseases and disorders and the purpose of the following sections is to describe the symptoms, causes and treatment of those most common. A beekeeper would be very unlucky to have personal experience of all of them, but if he keeps bees long enough, is likely to come across some.

Two general comments are worth making. First, the best way to recognize disease or abnormality, especially in the early stages, is to familiarize oneself with the appearance of normal healthy brood and adult bees. *Any* departure from this norm should be a matter for suspicion and investigation and perhaps advice and help from a more experienced beekeeper. Second, there has grown up an extraordinary and, to me, totally incomprehensible aura of secrecy about bee diseases, starting with Ministry of Agriculture officials who refuse to divulge the precise location of an outbreak of even the most serious disease, down to beekeepers themselves, some of whom whisper conspiratorially about cases of disease as though the beekeeper affected has something to be ashamed of. This is complete nonsense. There is no more stigma attaching to the beekeeper with an outbreak of bee disease than there would be if he were to catch cold. We need more openness and dissemination of information, not less; and less red tape, not more if we are to tackle these sporadic outbreaks properly.

American Foul Brood

The causative agent is a bacillus—*Bacillus larvae*. Spores of the pathogen, i.e. the resting stage, when fed by nurse bees to larvae with brood food, germinate within the body of the larvae into motile bacteria and develop at an immense rate. Soon after the cell is sealed and before pupation the larvae die. No more tissue is available for the bacteria to feed on but they do not die. Each bacterium forms itself into a spore. The spores are very resistant to disinfectants and dessication and can remain viable for many years.

When this happens an antibiotic is released with the result that no other micro-organisms can grow in the decomposing larval remains and these become an almost pure culture of AFB spores. We shall find that the opposite is true of EFB when we come to consider it and a characteristic of the two diseases is that microscopical examination of AFB remains show only AFB spores but EFB is always accompanied by other strains of bacteria.

The dead larvae eventually dry into blackish scales firmly attached to the bottom angles of the cells. The efforts of workers in cleaning out cells containing the remains of dead larvae or in robbing from colonies which have died out or have become seriously weakened by the disease result in their taking up thousands of spores and feeding them to previously uncontaminated larvae and so the cycle is repeated. The house bees doing the cleaning and the foragers doing the robbing will not, of course, be the nurse bees feeding larvae, but with the continual food-sharing within the hive, the spread of the spores is rapid. Adult bees seem to be unaffected—this is solely a disease of young brood. Infection usually takes place in the first three days of larval life. After this larvae acquire a certain amount of immunity.

The beekeeper may inadvertently help to spread the disease himself. If he fails to recognize early stages of an infection he may transfer combs from an infected colony to nuclei or other colonies and so spread the attack throughout his apiary.

Exposure of honey containers or extracted combs too may be a cause of spread. There have been too many cases of infection from drums of imported honey which have come from diseased sources. These are difficult, if not impossible, to prove but the weight of reasonable inference is too great to be ignored.

Symptoms to be looked for are a patchy appearance of the brood combs, cell cappings with a dark, greasy look and sunken and perforated cappings. As the cell contents start to dry and before reaching the dry scale stage the body mass of dead larvae develop an unpleasant glutinous consistency. If a matchstick is inserted into a suspect cell, twisted gently and withdrawn the material will come out as a glistening thread between matchstick and cell. This is the well known 'ropiness' test and is pretty well infallible.

There is no known cure and diseased colonies *must* be destroyed by fire and the hive and all ancillary equipment sterilized.

The colony should be killed after flying has ceased for the day. This is most easily done by completely blocking the entrance with sacking, clods of earth or something which will remain in place and pouring ½ pint (275ml) of petrol through the feedhole, closing it immediately. The petrol vapour will kill the colony in seconds. Dig a good hole and start a fire in it. Add every comb and all the dead bees. Scrape out the hive and add all the odd pieces of wax and propolis to the fire. This is a notifiable disease and the destruction should be done in the presence of a Ministry official. He will issue a certificate of destruction which will enable you to recoup some of your loss through insurance —if you are adequately covered against disease! So many of us are not.

The only really safe way to sterilize the hive and any other equipment is by scorching everything, inside and out with the flame of a blowlamp, paying careful attention to corners, nooks and crannies. It is a sad thing to destroy a colony in this way, but it is the only possible treatment.

European Foul Brood

The causative agent is again bacterial—*Streptococcus pluton*, nearly always accompanied by bacterium eurydice and/or bacillus alvei. Microscopical examination of the bacteria show variations from case to case. Only larvae are affected and they die when about four days old before the cells are sealed. *S. pluton* does not form resistant spores, but *B. alvei* does.

Affected larvae seem to wander about inside the cell instead of remaining coiled up at the base of the cell as normally. They die in twisted, unnatural positions—this is a characteristic of the disease. The dead larvae also make brown scales when dry but they are attached loosely to the cell walls and can be removed easily by house bees. The smell of the larval remains can be very unpleasant depending on which secondary bacilli are present— and there are others in addition to those listed above. This has given rise to the inclusion of the word 'foul' in the name of the disease. There has recently been a move to omit the word, which some people seem to find objectionable, but I have used the old nomenclature because it is in all the reference books and Ministry advisory leaflets.

Streptococcus pluton is fed to the larvae by nurse bees. It does not penetrate into the body mass, but remains in the gut and stomach and feeds on the brood food. It really kills the larva by starving it. House bees have no difficulty in clearing out dead larvae and it is often the case that a mild infection will be with a colony for years, flaring up and dying down spontaneously.

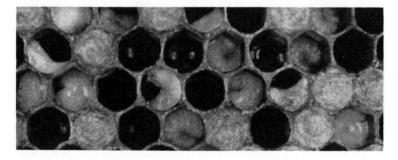

Figure 39.   European Foul Brood. Note twisted positions and 'melted' appearance of infected larvae.

Often there seems to be a seasonal outbreak of visible signs of the disease followed by an apparent recovery coinciding with the start of a nectar-flow, the disease reaching its lowest ebb as the nectar-flow reaches its peak.

There is some evidence that there are colonies where the disease is endemic, flaring up during the active summer months and dying away by August. Such colonies may produce a reasonable honey surplus year after year. It may well be much more common than is generally thought.

Minor cases can be treated with antibiotics, but the treatment I have seen applied by Ministry officials has, in many cases, resulted in the deaths of colonies and, at best, serious depletion of strength. There is, too, a danger that frequent use of antibiotics could quite easily result in the evolution of a resistant strain of the pathogen. Serious infestations are best destroyed by fire.

Sac Brood

This is not uncommon and is caused by a virus. Larvae die after the cells have been sealed and when they are at the pre-pupal stage. The outer cuticle becomes leathery and the body contents turn into a watery fluid. Eventually the dead larvae dry to a scale on the bottom angle of the cell with the darker head end curled upwards somewhat to give what has been described as the 'Chinese slipper' appearance. Attacks are noticed from early May to high summer and usually disappear by the end of the active season. The disease is not a killer of colonies, but it has a weakening effect and there is no doubt that honey production is adversely affected.

There is no curative treatment other than re-queening in severe cases.

Chalk Brood

This is a fungal infection of brood caused by the fungus *Pericystis apis*. Larvae die of the invasion after the cell has been sealed. The fungus continues to grow on the body tissues and the larva becomes a hard chalky white mass of mycelium bearing dark-grey or black fruiting bodies. Attacks seem to occur only when brood has been chilled by spreading the brood-nest and

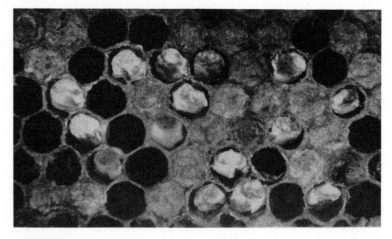

Figure 40.   Chalk Brood. Typical appearance of dead larvae.

similar manipulations. The precaution to be taken to prevent an outbreak is obvious.

Addled Brood

I do not think this is a disease in the accepted sense of the word. The cause is a genetic defect of the queen and is most common in inbred lines. Pupae die when nearly ready to emerge. They have very short, telescoped abdomens while heads and thoraces are normally developed. The cure is to re-queen.

Neglected Drone Brood

When a queen is beginning to run out of her supply of spermatozoa her production of queen substance usually begins to fail too and the bees set about superseding her. However, occasionally supersedure does not occur. The queen will then continue to lay unfertilized eggs in worker cells. These hatch into stunted drones and the nurse bees, possibly sensing that something is wrong, will neglect to feed and to look after the brood. The brood combs have an untidy appearance with scattered cells with raised drone cappings. The neglected larvae die and form dark masses at the bases of the cells. It is sometimes hard to distinguish between a drone-laying queen and the results of laying workers—unless the queen is seen of course. In either case the treatment is to re-queen with a mated laying queen or to unite what bees are left to a queenright colony if this seems worthwhile.

Chilled Brood

In spring, if a sudden cold spell follows a period of warm weather, the bees may move inwards towards the centre of the brood-nest and larvae of all ages occupying the outside cells will die of neglect. The dead larvae take on a blackish or dark grey appearance and lie in the bases of the cells from where the bees will eventually clear them out. The same conditions can result if the brood nest is spread too early in the season by inserting frames of drawn comb or foundation—or even when combs of young brood are exposed in a cold wind for too long during a prolonged examination. There is no 'cure' of course and the bees will clean and make use of the combs eventually when the larval remains dry out and shrivel.

Nosema Disease

This is probably the most widespread and weakening of the adult bee diseases. It is caused by a spore-forming protozoan parasite, *Nosema apis* which, in there productive stage, lives and multiplies in the cells lining the mid-gut. When the contents of these cells become exhausted the Nosema parasites cease active reproduction and form spores. The cell walls rupture and the spores accumulate in the mid gut, pass through the hind gut into the rectum and emerge in the faeces. The spores will remain dormant until they are licked up by other bees and so enter the mid-gut of a new host—the only environment in which they can develop.

Inside the spore there is a coiled filament and a twin nucleus. When a spore finds itself in a suitable situation the filament is extruded and the twin nucleus passes down the hollow centre of the filament into the protoplasm of the cell where it quickly grows and develops. The spores seem to have a hyaline casing or perhaps the contents have a high refractive index. In any event they are clearly visible in a suspension in water as clear rice-shaped grains with hard black outlines. A compound microscope with a good one-sixth objective will resolve them well.

In the healthy bee the cells lining the mid-gut are constantly sloughing off as part of the normal process of digestion, so it is possible for minor infections of Nosema to remain undetected. Some beekeepers go so far as to say that the disease is endemic in most colonies and only needs the right conditions—excessive damp or disturbance of the colony—for it to flare up. This may be going too far, but there certainly is evidence that there are many, many colonies with mild infections which show slight symptoms in the spring and apparently recover spontaneously in the summer.

Wet cool conditions and a poor honey harvest are often pre-cursors of an attack in the following spring. A succession of wet summers seems to have a cumulative effect.

Infected bees live only about half as long as healthy bees and their hypopharyngeal glands do not develop completely so that a badly affected colony will not be able to produce the expected supply of brood food.

Symptoms

Lightly infected colonies may show no clinical symptoms. Heavy infection will result in the dropping of patches of excreta on the combs and walls of the hive and surrounding vegetation with many 'crawlers', i.e. adult bees apparently unable to fly, and heaps of dead and dying bees outside the hive. However, all these signs may be symptoms of fermented stores, poisoning and other troubles, so care is needed before jumping to a con-clusion. The only sure diagnosis is the identification of the spores in the splashes of faeces by microscopical examination. If Nosema is the cause of the trouble quantities of spores will be seen in a watery suspension as already described.

Treatment

During late spring and early summer many infected bees are able to make cleansing flights away from the hive and as the old bees die and are replaced by young uninfected bees the colony appears to have made a complete recovery. However there will be a few odd spots of faeces in the hive to infect a small number of house bees and these will remain a focal point for a fresh infection.

The answer is hygiene and the regular replacement of old comb by clean foundation as a part of annual management. It is also possible to get a complete colony onto clean comb in the following way. Remove all unoccupied combs in the brood chamber and move the occupied combs to one side of the hive and add a division board. Add a second (clean) brood chamber with an equal number of frames of foundation to the combs below and position them vertically above the lower combs.

Finish with a division board, a cover board and a feeder of syrup. When there is a little brood in the new upper combs, which will have been drawn out, put a queen excluder between the boxes and make sure that the queen is in the top box.

When all the brood has emerged in the bottom box take off the top box and place it on a clean floor board on the old stand. Shake in the bees from the old brood chamber and continue to feed if in doubt about stores.

Take the old brood chamber and its combs away for sterilization. Fresh infection will come from contaminated combs, hive walls, etc. and not from the bees themselves.

If the colony is a large one you may not be able to find enough empty brood combs to carry out the above method. In this case, add a complete brood chamber of frames with foundation and then proceed as suggested. It is a help if you have a few frames of drawn comb, perhaps previously used as super combs, which you know to be clean, for inclusion with the frames of foundation. This would give the bees a good start.

After the bees have been transferred on to a clean comb in this way the addition of Fumidil B, an antibiotic derived from a mould, to the next autumn syrup feed is an effective prophylactic, but it is expensive.

Sterilization can be carried out by the fumes of acetic acid. Stack brood chambers containing suspect combs on an airtight base, one upon another and on the base and on each subsequent storey of combs lay an absorbent pad soaked with about $\frac{1}{4}$ pint (150ml) of 80 per cent solution of acetic acid. Be careful with the acid. It is very corrosive—even the fumes. Use rubber gloves and keep your face and eyes well clear. The fumes will attack metal, so remove metal ends if you use them. The fumes do not affect wax, honey or pollen. Leave the combs in your 'fume chamber' for at least a week and air when very thoroughly before using them again.

Amoeba Disease

This is also caused by a protozoan parasite, *Malpighamoeba mellificae* which invades the cells lining the malpighian tubules, multiply in them and form spores which pass down into the rectum and are voided with the faeces. The spores are spherical and quite unlike those of Nosema when examined microscopically. There is no known treatment—Fumidil is not effective. Fortunately, the disease is not very common in the United Kingdom. The best treatment is the hygienic replacement of combs as suggested under Nosema. Fumigation with acetic acid gives effective sterilization.

Braula Coeca

Braula is sometimes called the 'bee louse'. It is in fact a species of fly (*diptera*) which has lost its wings during the process of evolution. Adults cling to the backs of workers and queens (especially queens) and hang on round the constriction between the thorax and the abdomen. When the host bee is about to feed the braula moves up to the bee's mouthparts and partakes of whatever the bee is feeding on. Some observers have thought that the braula actually reaches inside the mouthparts and sucks up some of the glandular salivary secretion. They appear to do no harm to the bees, merely getting free travel and food, but one cannot help feeling that they must be a source of irritation. A puff of cigarette smoke never fails to make the braula drop off an infested queen.

Adult female braula lay their eggs just under the cappings of honey-comb and the hatched larvae are an unmitigated nuisance. They travel along eating pollen and wax as they go and make a network of unsightly tunnels.

Acarine

Acarine is caused by a mite, *Acarapis woodi*, which infest the breathing tubes (tracheae) leading from the first pair of thoracic spiracles of adult bees. The mites are only able to enter the tracheae of bees not older than five to six days. Bees older than this seem to be immune. Having crawled into the tracheae the adult mites soon begin to lay eggs. These hatch and go through a series of moults until they too become adult—all this within the body of the bee. The mites have a pair of fine stylets with which they pierce the walls of the tracheae and feed on the bees body fluids. As infestation gets crowded some of the adult mites leave the host bees through the spiracles and search for fresh young bees to cling to and so spread the infestation. This is most likely to occur during times of confinement within the hive when young and old bees are in close proximity.

Effects of Infestation
The thoracic tracheae affected are the two main trunks supplying air to the great indirect flight muscles and one would expect that the first thing to be affected, or at least impaired, would be the bee's ability to fly. Curiously this does not seem to be the case except in the most severe infestations and field observations have shown that bees with quite heavy infestations of acarine appear to be able to forage for pollen and nectar as well as normal healthy bees, although their life-span is much shortened. For this reason beekeepers will be unaware of mild cases of infestation and it is quite possible that acarine is much more widely spread than is generally supposed. After all, the success of a parasite depends on its *not* killing the host.

Dark brown patches appear on the walls of the infested tracheae. This was originally thought to be excreta from the mites, but closer investigation has shown that this is not so. Slight damage to the flight muscles has been observed and the nerves serving the wings can also be affected.

Acarine infestation is probably as rife in England and Wales as anywhere in the world but it has been reliably estimated that under 2 per cent of all colonies can be expected to show visible signs of sickness.

The visible signs are most likely to be found in March and April in heavily infested over-wintered colonies after a poor or a series of poor seasons. These colonies do not seem to develop vigorously like the others and later on masses of bees can be seen outside the front of the hive apparently unable to fly or to climb back. Abdomens may be distended and there will be signs of dysentery. Wings may be extended half-way with the fore and hind wings not coupled together and the fore wings may be trembling. None of these symptoms can be taken as a certain indication of acarine and may be due to other causes, but are clearly indicative that something is wrong. The only reliable diagnosis is the dissection and examination of the thoracic tracheae.

Diagnosis

Examination of the thoracic tracheae under low magnification will show the dark discolouration in the tracheae which is an infallible sign of the disease. A 15x hand aplanat lens will reveal the discolouration, but in the early stages of infestation there may not be much of this—merely a slight darkening—and the hand lens will not enable much to be seen. A 30x prismatic erecting microscope mounted on a stand (see under 'Microscopy') is much better and with it the mites can be clearly seen. This is an ideal, low-priced instrument for which the bee-keeper will find many uses.

The tools required are simple: a scalpel, or a sharp penknife; a wine bottle cork cut off at an angle of 45°; and a double needle. The latter can be made by binding together two sewing needles separated by about ¾ inch (2cm) of brass wire. The shaft of a pin will do with the head and point removed, but this is really a little too thin. A spot of solder will secure. You will also need a pair of tweezers (forceps) with fine points. Take a sample of about thirty bees in a matchbox from a suspect colony. Kill the bees by slipping into the matchbox a small piece of filter or blotting paper moistened with chloroform or entomologists 'killing fluid'. Petrol will kill the bees but will leave the dead bees curled up and with the muscles contracted. The other two fluids will leave the muscles relaxed.

Pin the bee to be examined, on its back, to the sloping cork by passing the double needle through the thorax at the bases of the legs. The purpose of the double needle now becomes clear. If a single needle were used the bee would revolve round on it.

Lay the scalpel (or the blade of your penknife) across the

thorax immediately behind the front legs. *Push* off the head and front legs. Push, do not cut downwards. The head and front legs will come off with a fine elastic white thread between them—this is the oesophagus. Disregard it. If you now look at your bee with your lens or dissecting microscope you will find that you have an opening in the thorax shaped rather like a broad figure eight. Through this you can see the main tracheae where they start to branch off. In a healthy bee they should be glistening and creamy white. Dark patches or oval-shaped mites will be a positive identification.

However, the parts of the tracheae you can see at this stage may look clean, but mites may be in those parts lying towards the spiracles and it is necessary for you to see the whole of the tracheae. You will see a kind of collar surrounding the hole in the thorax you have made. This is the tergite of the prothorax. The points of the collar nearly meet at the ventral side of the bee. Grip one of these points with your fine forceps and peel off the tergite. With a little practice this can be done in one gentle circular pull. If the collar breaks you will need to get it off in bits. Once the tergite is removed the whole of the tracheae will be exposed right down to, and including, the spiracle. A more complete examination can now be made.

Treatment

FROW TREATMENT is for use in late autumn and early spring. Robbing may be induced if it is applied during warm flying weather and this is why it is recommended for the cooler parts of the year. It will also disturb a tight cluster if applied in the depths of winter. Frow mixture consists of:

Nitrobenzene	2 parts
Safrol	1 part
Petrol	2 parts

This is a highly volatile, flammable and poisonous fluid and should be handled with great care. Safrol is now difficult to obtain and the following modified Frow mixture may be used:

Nitrobenzene	6 parts
Methyl salicilate	2 parts
Petrol (or Ligroin)	5 parts

This gives good results but is probably not as effective as the original Frow recipe.

30 minims (1.8ml) of the mixture is poured onto an absorbent pad and this is inverted over the feed hole of the affected colony. The pad should be covered with a tin lid to prevent wasteful evaporation upwards. The dose is repeated every other day until seven doses have been given. While the treatment is being carried out the hive entrance should be reduced to one bee-space.

In the case of out-apiaries or when repeated visits are inconvenient, one single large dose can be given. A dose of 4.5ml is given as described and the pad left for fourteen days. The single-dose method has its dangers. If the weather is mild the bees, or a large proportion of them may be killed by the heavy concentration of fumes. The method is best used only in cold weather.

It is only fair to add that some research workers have shown that infested colonies treated are more likely to die out in winter than they would have done if not treated. As in other branches of beekeeping caution is needed with medication.

FOLBEX STRIPS are pieces of card about four inches (10cm) long impregnated with an acaricide, chlorobenzilate, and are best used in early summer before the main nectar flow is expected.

An empty shallow super is placed on the hive below the crown board and in the evening when flying has ceased the entrance is blocked. A strip of Folbex is pinned to a strip of wood. The bottom edge of the Folbex strip is lit and the flame blown out allowing it to smoulder. The Folbex strip is inserted through the feed hole so that it hangs down in the empty super. The hive is left for an hour and the Folbex then removed.

Both Frow mixture and Folbex strips can be obtained from the beekeeping appliance manufacturers, the former in capsules containing suitably measured quantities.

The vapour of methyl salycilate alone has been used. It is true that this will kill the mites, but the strength required is very close to that which will kill the bees too. It also damages unsealed brood.

I cannot help having serious misgivings about the wisdom of dosing diseased and enfeebled colonies, whether they are suffering from acarine or any other disease. I have doubts whether such colonies ever regain their proper vigour, at least not for some years and then only after re-queening, and there is too, a real danger of breeding resistant pathogens. Nobody likes killing bees, but it is at least worth considering whether rigourous culling coupled with a programme of raising nuclei from stocks known to be strong and productive would be a better proposition. I am sure that indiscriminate dosing with all sorts of nostrums as prophylactics is to be deprecated. If we give nature a chance, she nearly always repays us handsomely.

Isle of Wight Disease

A word about this scourge which swept through the country in the early years of this century decimating the bee population. A legend has grown up that in the disease was a flare up of acarine infection. This is based on very slender evidence. It seems arguable that it was not due to a single cause but to a number of agents, of which bacterial and virus infections and bad beekeeping methods were not insignificant, in addition to the primary infestation of acarine.

1906, probably the peak year for the disease, started disastrously for agriculture with frost, hail and snow in April and May. A very hot spring followed. It is reported that among the 'cures' used by beekeepers in combating Isle of Wight disease were beef tea, carbolic acid, formalin, napthol, izal, alcohol, onions, sour milk, vinegar, salt and jalap! No wonder the bees died.

The disease left the United Kingdom denuded of bees and this, coupled with the importation of thousands of bees—queens and package bees—from abroad saw the end of the old British Black bee, although it has been claimed that pure strains did survive in isolated pockets. In view of what we know about the mating habits of the honeybee it is hard to believe that there can be any pure strains left now.

Paralysis

In the past the presence of obviously sick bees, unable to fly, abdomens distended and with trembling wings was put down to acarine, Nosema disease or a combination of both. This was in the days when the light microscope was the ultimate tool. Nowadays the electron microscope is a common research tool and the techniques of serology have been applied to bee diseases. All beekeepers owe a great debt of gratitude to Dr L. Bailey at Rothamsted, and others, whose painstaking work has thrown new light on the problem.

It now seems most likely that the symptoms described at the head of this section are often due to infection by one or more viruses. A number of these has now been isolated and identified. As with the common cold or influenza in man (both virus diseases), the effects of viruses on bees can be mild or very serious. Transmission of the diseases from individual to individual is not yet fully understood. It may be that there are genetic factors which tend to make some strains of bees more susceptible than others.

In vertebrates, mild attacks of virus diseases are common and often remain undetected because the symptoms are sub-clinical. Might it not be the same with bees? As mild attacks frequently give rise to the production of antibodies in the affected individual and resultant resistance, even immunity from further attacks, could it be possible that something of a similar nature takes place in bees? These are mere speculations, but research continues and beekeepers are extremely lucky to have people like Dr Bailey at its head.

Varroasis

This is caused by infestation of a mite belonging to the spider family (*arachnidae*). They have eight legs as compared with the six of *Braula coeca*. They live by sucking the blood of adult bees, larvae and pupae and spend a good part of their lives within the brood combs on the body of juvenile bees.

Infected bees are much weakened, their life spans are halved, the colony dwindles and eventually dies. The European bee, *Apis mellifera*, has no natural resistance and succumbs easily.

The only method of transmission is from infected bee to healthy bee. The mites cannot survive on their own away from bees for more than nine days.

The disease has been known for many years in the Far East, where it is endemic, and the local bee, *Apis cerana*, seems to

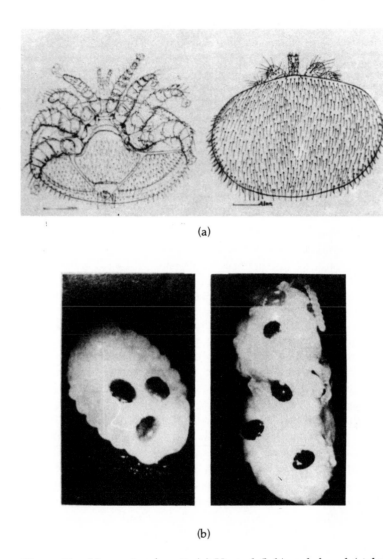

(a)

(b)

Figure 41.   Varroa jacobsonii. (a) Ventral (left) and dorsal (right) view of adult mite. (b) Mites infesting larvae and pupae while still in cells.

have developed a degree of tolerance. In recent years the disease has started to migrate westwards across Asia and Russia into Europe and has now been positively identified in West Germany. There are reliable observers who have said that it is in France too. So far it has not crossed the *cordon sanitaire* of the English Channel and, at the moment we are free of it. Let us keep it so. The mite can only be introduced into the United Kingdom on the bodies of bees and this is why beekeepers, individually and in associations have been pressing for legal controls on the import of bees from abroad. These efforts have been partially successful. The import of bees from a long list of specified countries is now prohibited and there are licence requirements from others.

Any country at present free from this disease can remain so by the simple expedient of not importing any bees at all. Any beekeeper who buys imported queens with attendant workers ought to be aware of the potential risk.

Spray and Pesticides

Anyone who has anything at all to do with food production will know the enormous damage done to food production done by pests, and this at a time when world food production is inadequate to feed the total population. Even on a miniscule scale the trouble is brought home to the gardener with two or three rows of broad beans with an infestation of black fly.

The farmer and grower tends to see the problem as one of finding a pesticide which, ideally, will kill 100 per cent of the pest and leave the crop intact. The environmentalist, and I think we can include the beekeeper in this group, will want to put an end to the use of all pesticides and herbicides.

Somehow or other we have to find a compromise between these two fundamentally opposed points of view. I do not think that the environmentalist's extreme policy of no pest control at all is tenable.

Nature *does* need a helping hand and we need only think of locusts, cotton boll weevil, colorado beetle and aphids to see that with an escalating world population some control is essential, unless, that is, we are content to let pestilence, war or some natural disaster reduce the level of human population very dramatically indeed.

On the other hand, there is no doubt in my mind that for far too long we have indulged in an orgy of overkill, using the wrong pesticides, at the wrong strength and at the wrong times. It may be that a large part of the blame lies with our system of marketing. The pesticide salesman's job is to sell his wares in the greatest possible quantities to as many outlets as he can find. In too many cases his help is sought by the farmer in spite of the fact that there is available an excellent source of advice and information in the shape of the Government-sponsored Agricultural Development and Advisory Service (ADAS) whose branches are staffed by competent and unbiassed officers. Rachel Carson's *Silent Spring* (Penguin, 1970) is not a happy book, but it might be a good idea if it were required reading for all agriculturists.

How can the situation be ameliorated? In the first place all beekeepers should make themselves known to neighbouring farmers and let them know where hives are situated. Farmers are usually reasonable people and are aware of the considerable good bees do for their crops by way of pollination. They will be prepared to warn of impending spray programmes. Beekeeping Associations can help too in running 'early warning' schemes under which nominated liason officers are notified of proposed spraying and disseminate the information among members likely to be affected. Such schemes have limited use I think. If a beekeeper is notified of coming spraying what is he to do? He cannot shut his bees up, certainly not in warm weather, without as many bees dying of suffocation as would have been affected by the spray. He could move his bees at night to a safe situation at least two miles away but is this practicable? To move even a small apiary of ten or fifteen hives in the active season is a formidable proposition. The man with a flat bed truck who specializes in migratory beekeeping or pollination could do it but would still need 48 hours notice. It seems unlikely that the small beekeeper, probably out at work during the day, could manage it.

The problem could be eased by using the pesticides known to be the least toxic to beneficial insects. It is not only honeybees that are affected. Bumbles and solitary bees, ladybirds, lacewings and parasitic wasps all fall victims and some of these are natural predators on the pests we seek to control.

In their cool, calculating way, scientists have evolved a scale of evaluating the toxicity of pesticides. It is the measured dose that will be lethal to bees. The table below is a list of the more common pesticides with a note of their comparative toxicity. The amounts required to kill are extremely small. A microgram is a millionth part of a gramme or one twenty-five-millionth of an ounce. It will be noted that pirimicarb-based pesticides are the least harmful, but it does not follow that this should be used in all cases. Different pesticides are more effective against different pests, but a choice of varying toxicity does exist and in too many instances the recommended strength of the spray is exceeded.

Table 8.   Comparative Toxicity of Liquid Pesticides to Worker Bees

Pesticide	Microgram per bee
fenitrothion	0.018
demeton S methyl sulphone	0.020
azinphos methyl	0.063
dimethoate	0.12
gamma HCH	0.20
demeton S methyl	0.26
malathion	0.27
phorate	0.32
oxydemeton methyl	0.54
carbaryl	1.3
disulfoton	4.3
menazon	4.3
endosulfan	7.1
pirimicarb	54.

The layman does not know which pesticide does what but ADAS does know and is there to help.

The time of application is important too. More bees are flying between 7 a.m. and 8 p.m. than at other times of the day and spray applied earlier or later than these times is less likely to harm large numbers of bees. Late evening spraying will also have a chance to dry off overnight and this too will be less harmful.

Application of spray by tractor confines the spray to the crop for which it is intended. Aerial application contaminates an enormous volume of air above the crop and even a slight breeze results in drifting. There have been horrifying cases of contamination of cattle, pets and even children by drift. And what happens when the pilot forgets to shut of his apparatus for a minute or so?

To summarize, progress might be made on the following lines:

—More co-operation between beekeeper and farmer.
—Consultation with competent advisers before spraying.
—Use of the least toxic spray practicable.
—Application of spray by tractor drawn apparatus.
—Use of pesticide in granular form whenever possible.

Wax Moth

There are two species of wax moth which are a nuisance and can do considerable damage to stored comb. Both are particularly fond of comb in which brood has been raised. In addition to the wax, both like the old cocoons and the faecal matter remaining in the cells. Some strains of bee even tolerate the presence of wax moth larvae in the brood chamber. They can easily be seen as can their silk-lined tunnels.

The larvae of the Greater Wax Moth (*Galleria mellonella*) when fully grown are about 1¼ inches (32mm) long, pale grey with a brownish head. The Lesser Wax Moth larvae (*Achroia grisella*) is much smaller, about ⅝ inch (16mm) long. Both can reduce a good comb to little more than a unpleasant mass of silk tunnels and dust in a matter of two or three weeks. In addition to damage to comb the Greater larvae can do considerable damage to the woodwork in the hive. When ready to pupate the larvae dig out little troughs in the wood in which to spin their cocoons and pupate. They seem to be gregarious to some extent so, although the damage is seldom structurally weakening, it is most unsightly.

Apart from the occasional Greater larva found in a brood chamber, most of the attacks take place in stored comb in the inactive part of the season. The best treatment is to store comb in stacked supers with a generous handful of PDB (paradichlor benzene) on paper between each layer. Storing in a cold situation open to frost is a good idea too.

It is sometimes claimed that supers stored still wet from the extractor will not suffer attacks from wax moth. I think this may be so but, having tried it, I find that one of two things is likely to happen. Either the honey crystallizes in the combs and these crystals form nuclei for crystallization of next year's crop in the comb or it absorbs water from the atmosphere and starts to ferment.

I think dry storage with PDB is the most satisfactory way to deal with it.

14

Microscopy and Dissection

Beekeepers and amateur entomologists normally do not have access to well-equipped laboratories, usually have little experience of refined techniques and therefore limited facilities for practical work in honeybee anatomy. The purpose of this chapter is to show that by using simple apparatus, much of it improvized or home-made, a whole new world becomes available to the beekeeper. The second requirement is a simple guide to the various systems of work.

I had always been an amateur naturalist. Perhaps that is too grand a word, but since schooldays I have been interested in living things, particularly insects, and had been used to using one of the cheap compound microscopes which, between the wars, could be picked up for a few pounds in secondhand shops. And very good they were too. Objects which needed no preparation gave no trouble. For instance, the hundreds of forms of microscopic pond-life provided years of study and pleasure. Nothing more than placing a drop of pond water on a slide was required.

However, when I turned to my insects I was immediately in difficulties. The elementary dissecting methods most of us have experienced in biology classes at school were of little help. Pinning out a large animal such as a frog, rat or rabbit and cutting downwards was suitable at the time I suppose, but certainly not when I came to my bees. They were too small to be pinned out, the external chitinous skeleton was difficult to tackle with the normal dissecting instruments and when finally opened, the soft internal tissues collapsed into a hopeless tangle of unidentifiable bits and pieces. Magnification too was a problem. My compound microscope was useless—the image was reversed and inverted. I had not at that time devised a workable alternative.

About this time I was fortunate to meet the late H. A. Dade at the Queckett Microscopical Club and subsequently attended his practical extra-mural courses at London University. I quickly discovered that my approach to problems and my method of tackling them had been entirely wrong. It was necessary to start again right from scratch. Much of what follows is based on Dade's teaching (for which I shall be eternally grateful) with modifications of my own. These not necessarily improvements, but perhaps alterations made to suit my own inadequacies. Dade was a perfectionist and his dissections were quite beautiful. I could not hope to come anywhere near his expertise; nevertheless his help and guidance enabled me to pursue my own studies with far fewer failures and frustrations than before and with considerable enjoyment.

Magnification

The purpose of magnification is to bring an object being examined closer to the eye, as it were. We have all known short-sighted people who, when they want to look closely at some small object, say a postage stamp, remove their spectacles and bring the stamp to within an inch or so of their eyes. They can then see the finest detail. People with normal sight cannot do this and cannot focus without strain at distances closer than about ten inches. Interposing a single lens between the object and the eye brings the object up. Better results are obtained by using a combination of two lenses—either an aplanat or a doublet (see *Glossary*). In the latter the spherical and colour aberrations common to single lenses are reduced.

These pocket lenses have working distances of one or two inches. As dissecting tools this leaves very little working distance for handling instruments and to use them on a table or bench leads to a cramped position for the operator. Still, a doublet can be mounted on some kind of sliding support, e.g. a knitting needle in a large cork, or on a hinged arm (I have used this with some success—and still do so).

As we have already noted, a compound microscope is unsuitable for dissection purposes because of the inversion of the image. Tools appear to move in the opposite direction to that actually taken, and it is not possible to work under these conditions; in any case the magnification is far too great.

What is needed is a prismatic microscope which will correct the inversion, provide a working distance between objective and object of not less than three inches (7.5cm) and provide a magnification of 20 to 30 diameters. Commercial instruments such as the *Greenhough* and the bench models made by Bakers and Leitz are all good. However, they seldom come on the secondhand market, and when they do they are expensive. New models by modern manufacturers are, of course, excellent but are too dear to be considered here. I need hardly add that the purchase of a commercially produced binocular dissecting microscope would be a first class investment—if the cash is available!

Constructing a Prismatic Microscope

This difficulty need not deter the student however, for a very servicable monocular instrument can be constructed quite cheaply from easily obtained materials.

The basic requirement is a prismatic erecting system—either two prisms or a Swan cube which is really two prisms cemented together. A fruitful source is the barrel of an old pair of binoculars—perhaps the undamaged half of a broken pair. I advise constant browsing in junk shops or the barrows of miscellaneous bric-a-brac in street markets. I was lucky enough to find the sighting telescope (of some ancient artillery piece I believe) which was small and adaptable. As a last resort try mounting two prisms in some kind of metal box. This is fairly easy to do, but not so easy to make a neat job of.

The eye and field lenses of your half-binocular should be discarded although the metal mounts may be useful, especially if of brass and therefore solderable. At one end solder, or fix with jewellers epoxy resin adhesive, a short brass or copper tube of sufficient internal bore to take a standard microscope eyepiece. Beg, buy, or obtain by nefarious means an old-fashioned brass microscope objective box. Look inside the screw cap. The kind you want has an internal female thread into which the objective is normally screwed. Cut out and file the centre of this lid and then solder it to the bottom end of your half-binocular. Remember you have glass prisms inside so be careful of heat. You now have an erecting system into the top end of which you can drop a microscope eyepiece—I suggest a 6x, and into the other end you can screw a microscope objective—a 2 or 3-inch one give magnifications of 20 to 30 diameters.

You will now need to mount your masterpiece on a firm base with some kind of focusing coupling. Various methods spring to mind, e.g. a sliding pillar with a collar, parallel metal arms. Every household should have a scrap box to be scrabbled through. I unearthed an ancient focusing mount from a Victorian magic lantern. This has a tube sliding in the mount plus a very small rack and pinion. The lenses you will be using will give a good depth of focus, so you will not need anything too fine.

Make your base large enough to rest your hands on when working and mount your microscope projecting out over the base as far as convenient. It is sometimes useful to put a whole comb on it for microscopical examination.

Illumination

It is necessary to illuminate your dissection with an intense pool of light. Remember that intensity of light varies inversely with the square of the distance (or of the magnification). I find the most useful set up is a 40 watt car headlamp bulb run off a transformer. Either a lens or a concave mirror can be used to throw an image of the filament onto the work. I use a mirror— again found on the ubiquitous junk stall. It is actually part of a discarded Aldis signalling lamp. It has a diameter of 3 inches (7.5cm). The lamp housing should have a hood to prevent the

Figure 42.   Home-made dissecting microscope and lamp.

light shining into the operators eyes and should be mounted on a pillar and be adjustable for height and angle. I find it most comfortable if the light falls at an angle of about 45 degrees and is somewhat to my left. Metal rods, clamps odd nuts and bolts and cleaned food tins can all be pressed into service.

The Tools of the Trade

These are neither complicated, nor with two exceptions, expensive. You will need at least the following:

1. Two needles mounted in wooden holders. (Choose fairly stiff needles for this. The fine needles used for fine needle-craft are too flexible.)

2. A *very* sharp small surgeons scalpel or similar. An eye knife is suitable, but I prefer a Swan Morton universal handle. This takes renewable blades of which the best are straight No. 11 and curved No. 12.
3. A piece of brass wire about $\frac{3}{32}$ inch (2mm) thick about 6 inches (15cm) long with $\frac{3}{4}$ inch (19mm) at one end turned at a right angle.
4. A pipette with a rubber teat.
5. A pair of blunt forceps—any old kind will do.
6. A pair of very fine scissors which are very thin in profile. The scissors sold with students biology sets are quite useless and only suitable for the coarsest dissection. A good pair will last a lifetime if looked after and the joy in using them is unbelievable.
7. A pair of fine watchmakers forceps. These are surprisingly dear, but you will find that the points meet exactly with no crossing or overlapping. These again will last a lifetime.

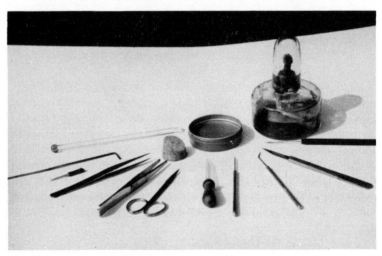

Figure 43.   Dissecting tools.

As your work proceeds you will certainly acquire odd imple-
ments which take your fancy or which you make up to suit your
own needs. A needle with the extreme point turned over into a
tiny hook or a needle hammered flat when cherry red hot and
then with the distal $\frac{1}{2}$ inch (1cm) bent at an angle are both useful
for lifting delicate structures or for reaching under a coiled
ventriculus etc. I always find a use for very fine pipettes.

Basic Techniques of Dissection

On reflection I do not think that my earlier plan of 'learning as
you go along' is a good one. I now think that before undertaking
practical dissection it is best to get familiar, in theory, with the
general outline of the bees anatomy both external and internal.
This need only be an outline. As practical work proceeds
referring back to books will prove useful. When opened up, the
bee's abdomen for instance is a confused jumble of parts of
various systems. When familiarity with the general appearance
is achieved, it is probably best to spend a session or so concentrating on one particular system, ignoring the parts of all others.

Having made this point, it is still a good idea to start with one
or two general dissections in order to be able to recognize the
various parts and their functions.

Before starting dissection the bee's body must be fixed firmly,
but without squashing or distorting it, so as to leave both hands
free for manipulation. This is best achieved by partially
embedding the bee in a dissecting dish containing the substance
we are all familiar with—beeswax.

Get a few tins about $2\frac{1}{2}$-3 inches (65-75mm) in diameter
(empty shoe polish tins, typewriter ribbon containers, for
example). The tins should be shallow. Fill them with molten
beeswax to within about $\frac{1}{4}$ inch (6mm) of the rims. Allow to
cool.

Take a freshly-killed bee and cut off the legs and wings with a
small pair of scissors (*NOT* your best dissecting scissors). If the
bee has died with its proboscis extended, cut this off too.

Your bee is now ready for mounting in your dissecting dish.
Heat the bent end of the brass wire tool and with it melt a little
pool of wax in the dish. Hold the bee by the thorax, dorsal side
uppermost with the blunt forceps and lower it into the wax pool.

It usually helps if the tip of the abdomen is lowered first, followed by the rest of the body, drawing it very gently forward. In this way the abdomen is slightly extended by stretching the inter-segmental membranes. Reverse the brass wire and press very gently on the bee's thorax with the straight (cool) end. Release and remove forceps. The object is to embed the bee in wax to about half its own depth. You may find it necessary to draw up a little more wax, freshly melted for additional security.

Immediately flood the dish with a 50 per cent alcohol solution. The great enemy of dissecting under water is the mass of air bubbles which collect on the bee's hairy body. A 50 per cent alcohol solution will obviate this trouble and will 'wet' your specimen thoroughly.

Place the dissecting dish on the baseplate of your dissecting microscope, switch on the spot lamp and focus this on your specimen.

In order to make a general dissection of, say, the abdomen you will have to remove the top of the chitinous external skeleton without disturbing or injuring the contents. Steady the dissecting dish in the left hand encircling it with the finger and thumb. Rest the fine scissors against the thumb and insert one point of the scissors just behind the rearmost visible abdominal tergite (usually the fifth) and cut right round the abdomen taking small snips and trying not to thrust the point in deeply. Keep the inside blade of the scissors, as nearly as possible, parallel with the body wall. As the angle of cut alters, do not move the right hand to compensate. Keep this where it is and move the dish round as required. Continue round until the point of your first insertion is reached. The 'lid' you have made can now be lifted off using the needles, taking with it the dorsal diaphragm and the heart. The lid will be anchored down possibly by the fine network of tracheoles that appear as delicate silver threads. These can be freed with the needles.

Dissecting Thorax

A different technique is needed to expose the contents of the thorax. This is almost entirely filled with the fibres of the flight muscles in four bundles. Two are attached fore and aft and the other two from floor to roof of the thorax. In a freshly-killed bee these fibres are gelatinous and difficult, if not impossible, to handle neatly. However if the bee is steeped in a hardening agent (e.g. formol alcohol) for a few days the muscle fibres will become tough and much easier to cut.

Fix a 'pickled' bee in a wax dissecting dish as described earlier. Scissors are unsuitable for a thoracic dissection. Use a sharp scalpel. (I use a Swan Morton handle with a straight No. 11 blade.) Cut a ring round the thorax towards the outside edge as viewed from above and also a straight run down a median line from front to back, using short nicks of the very tip of the scalpel blade cutting outwards. Having completed a circle you will find that the roof will still not come off because of the muscle attachments. Insert the scalpel blade horizontally and sever the muscles from the roof a little at a time. Joint use of scalpel and fine forceps will enable you to lift off half the roof, i.e. up to the median cut you have made. Remove the second half in the same way. If possible cut down the walls of the thorax to about the level of the wing roots.

USEFUL FORMULAS

Killing

The vapours of many volatile oils will kill bees and other insects, e.g. the vapour from say $\frac{1}{2}$ pint (275ml) of petrol will rapidly kill a colony of bees if this is required.

Unfortunately, most of the petro-chemical oils kill with the muscles contracted and this is undesirable from the point of view of dissection. Chloroform and ethyl acetate are much better killing agents. A few drops of ether on a small piece of filter paper inserted into a matchbox full of bees will quieten at once and kill in a matter of seconds with the muscles relaxed and abdomens extended.

Local pharmacists are sometimes reluctant to sell chloroform other than to medical practitioners. However the killing fluid sold by specialist shops works very well. Finely-chopped laurel leaves (*Prunus laurocerasus*) fairly tightly packed in a tin or wide mouthed bottle will make a good killing bottle.

Fixation

The process of putrefaction (by bacterial action) and autolysis (destruction of cell contents by its own serum) commences *immediately* on the death of the individual. Tissues will, therefore, alter rapidly and will shrink and become distorted by dehydration. In order to arrest these processes biological material should be subjected to fixation, the objects of which are:

1. To fix the tissues in as life-like a condition as possible.
2. To harden the tissues so that they can be handled and subjected to further treatment.

After fixation tissues should be preserved for future use in a suitable medium.

It is often thought that alcohol is a good preservative and fixative, but this is not so. Used alone it does not penetrate quickly enough and renders tissues much too brittle. There is probably no single substance which is completely satisfactory as a fixative, but two materials make good the deficiencies of alcohol. These are formalin and acetic acid.

I have found the two following satisfactory:

1. *Formol Alcohol*

95 per cent alcohol (Industrial Methylated Spirit)	70 parts
water	25 parts
formalin	5 parts

2. *Carls Solution*

95 per cent alcohol	17 parts
formalin	6 parts
acetic acid	2 parts
water	28 parts

Dry Cells

A number of parts of the exoskeleton are best mounted dry. If parts selected for this method of preservation contain soft tissues, e.g. the head, they should be exposed to the vapour of formaldehyde for a few days in a covered dish (a piece of cotton wool soaked with formalin will provide the vapour). This discourages the formation of moulds. Do not try to seal the mount otherwise condensation and the growth of micro-fungi are almost certain to occur. In addition, the contents of an unsealed mount can be renewed with little difficulty when it becomes necessary.

Suitable parts for this treatment are:

— The heads of all three casts mounted to show all aspects.
— Legs—inside and outside aspects.
— Thorax mounted to show lateral aspect with wings cut down almost to the roots.
— Mandibles—all casts.

Method

Cut slips of ⅛ inch (3mm) plywood exactly the size of a standard micro-slide (1 in × 3 ins/25mm × 75mm). In the centre cut a ¾-inch (19mm) round hole. Smooth with fine glasspaper. Gum or glue one side of the plywood strip and place on a piece of white card, about the thickness of a visiting card. Leave overnight under weight and then trim with a razor blade or very sharp knife precisely to the plywood slip.

Using similar white card, make a shallow tray with two long sides which will cover the bottom of the plywood strip, extend up the sides as far as a micro-slide placed upon it. If the positions of the creases are lightly scored with a sharp knife a neater fold will result.

Bind the micro-slide to the card tray with wide passe-partout (a paper tape gummed on both sides) or some similar material, passing underneath, up the sides and over the edge of the glass slide. Leave the plywood slip in place as you do this.

You will now have made a small-glass topped drawer the sliding part of which is the plywood slip. Mount your specimens on the card at the base of the hole with a strong synthetic glue.

Mounting a Cornea

An interesting dry mount can be made of the cornea of a compound eye. Cut a slice off a compound eye with a scalpel. Invert in a watch glass in alcohol and wash out the soft parts with a camel hair brush—but gently! The cornea is convex so it will not be possible to flatten it completely, but a few small snips round the periphery with scissors will help. Place between two micro-slides separated by two pieces of fairly thick card on either side of the specimen, bind with cotton and immerse in alcohol for a day or so. Cut a card ring, as narrow as you are able, the size of your cover glass. Remove a small wedge from the ring for ventilation. Glue the ring to a micro-slide with adhesive. Secure the cornea to the slide within the ring with two or three *small* dabs of adhesive *at the edges.*

Take great care that the adhesive does not run under the cornea. If it does your mount is ruined. Success depends on the refractive indices between the cornea and air. Apply adhesive to the ring and lower on a cover glass. Allow to set. When quite set place on your microscope stage under a $\frac{1}{6}$ inch (4mm) objective. The cornea will be seen to consist of many hexagonal facets with stiff hairs between. Each facet is in fact a separate lens, and this can be demonstrated as follows:

Place a black card with a simple design cut out—a cross or a 'V' in front of the microscope lamp. Rack the objective up a little until the image blurs and at the same time rack down the substage condenser. With a little fiddling with both controls images of the crosses will come into focus (one for each facet). It should be noted, however, that this does *not* prove that the bee sees a large number of separate images. In life each facet is backed by a crystalline rod (rhabdon) and each rhabdon is surrounded by an opaque pigment. The overall effect is a kind of mosaic of dots of light of varying intensities on the retinal nerve cells.

Preparing Parts of Exoskeleton for Mounting

Most parts of the exoskeleton are too deeply coloured to make satisfactory slides. The answer is bleaching. Any of the common domestic bleaches will do this quite satisfactorily. The parts to be bleached are immersed in the fluid in a suitable receptacle. (I find the opaque white pots sold containing *paté* are useful for this because the process of bleaching can be watched.) The ideal is to bleach to a light tan colour. When this point is reached replace the bleaching fluid with water and change the water three or four times to wash out the bleach.

After bleaching it is an advantage to dissolve out any soft tissues—the exoskeleton is what we are after.

Transfer the specimens after the last washing into a 10 per cent solution of caustic potash in which they can safely be left for two days. After this, wash very thoroughly with five or six changes of clean water. This process is known as 'macerating' and is intended to remove all the soft internal tissues.

Dehydrate in three changes of alcohol and, finally, two changes of cellosolve (a substitute for absolute alcohol). Clear in clove oil to remove cellosolve. Wash out clove oil with xylol. Mount on a micro-slide in Canada balsam. (Canada balsam unfortunately is now very expensive, but I do not believe that any satisfactory substitute has been found yet.) Try to arrange your specimen in the balsam with warm needles and then apply a cover glass. Hold this down with light pressure and place on one side to set. A suitable clip can be made from a twisted paper clip—the amount of pressure can be readily adjusted.

Suitable subjects are:

— Legs.
— Spiracles—cut out part of an abdominal tergite to include a spiracle.
— Proboscis. Remove the head from a freshly killed bee and dissect out the proboscis including the cardines (hinges) in your specimen. Place in water in a watch glass and tear the membrane which unites all the parts nearest the head. This will enable you to spread the proboscis out flat.

   Lay out and arrange on a micro-slide or a small piece of glass. Cover with a second slide with slips of thin card on either side of the specimen and tie the two together with cotton. Immerse in alcohol for two days. The proboscis will dehydrate and harden in this time and will remain in the position in which you set it. In a watch glass pass through cellosolve, clove oil, and xylene and mount in balsam. Apply light pressure with clip until the balsam hardens.
— Antennae. Bleach, macerate in caustic potash, dehydrate, pass through cellosolve etc. as above and mount.
— Wings. Mount in pairs under pressure with clip. No bleaching or maceration necessary, but dehydrate in cellosolve etc.
— Sting apparatus. Dissect out from a freshly killed bee. Do not bleach, but macerate in caustic potash. Remove

proctiger (the last abdominal segment which contains the anus), wash and dehydrate. Mount in balsam.

Such a mount will display the quadrate and triangular plates which would be obscured if maceration had not been done.

Glycerine Jelly

Glycerine jelly is a mountant for micro-slide specimens. It can be constituted as follows:

1. gelatin   1 part by weight
   distilled water   6 parts by weight
       Leave for at least 2 hours and add
   glycerine   7 parts by weight
       For every 4 oz (100g) of the above mixture add
   thymol   a *few* crystals—not too much or the jelly
                                                               will become cloudy
       Warm in a water bath until gelatin is completely melted.

2. gelatin   50g
   distilled water   175ml
   glycerin   150ml
   phenol   7g
       Soak the gelatin in the water for 2 to 4 hours. Add glycerin and phenol and warm the mixture over a water bath for about 15 minutes, stirring frequently. Store in a receptacle with an air-tight lid.

Both the above will keep indefinitely if lids are kept screwed down tightly. To each add only a *few* drops of basic fuchsin (10 per cent aqueous solution). On the whole I think I prefer the formula using thymol because I believe phenol causes fading of stains after a time.

Smears

After making and drying smears a good mountant is ordinary *Durofix*. Make a thick line of *Durofix* at one end of the slide and spread this over the smear with one smooth even movement of a finger. No cover glass is needed.

Appendix 1

The Composition of Honey

Constituent	Percentage content
Water	17.2
Fructose	38.19
Glucose	31.28
Sucrose	1.31
Maltose	7.31
Higher sugars	1.5
Acids (glutonic, citric, malic, succinic, formic, acetic, butyric, lactic, pyroglutamic, amino acids)	0.5
Ash (potasium, sodium, calcium, magnesium, chlorides, sulphates, phosphates)	
Pigments (carotene, chlorophyll, etc.)	
Aromatic substances (terpenes, aldehydes, alcohols, esters)	
Sugar alcohols (mannitol, dulcitol)	2.21
Tannin	
Enzymes (invertase, diastase, catalase, phosphatase)	
Inhibine	
Vitamins (thiamine, riboflavine, nicotinic acid, vitamin K, folic acid, biotin, pyridoxine)	
Pollen grains	

The figures given are for an average good honey. They will vary with the season and the plant source. (Figures condensed from the *Hive and the Honeybee*, ed., Roy A. Grout, published by Dadant & Sons, Hamilton, Illinois).

Appendix 2

Important Plants

Below is a list of the more important plants in the U.K. which are visited by bees for pollen and/or nectar (plants marked * yield pollen only). Dates are approximate and vary according to long-term weather trends. They are for average weather in the south of England. The list is not meant to be exhaustive and the honey from some plants (e.g. privet and ragwort) is unpleasant.

Major Honey Plants

April	*May*	*June*
Plum	Apple	Blackberry
Damson	Dandelion	Field bean
Cherry	Hawthorn (not always)	Raspberry
Pear	Sycamore	Sainfoin
		White clover

July	*August*	*Sept./October*
Bell heather	Bell heather	Ling heather
Blackberry	Blackberry	Mustard
Limes	Ling heather	
White clover	Red clover	
Willowherb	Willowherb	

Plants of Lesser Importance

Feb./March	April	May
*Alder	Almond	*Beech
Almond	*Ash	Bilberry
Butter burr	Berberis	Bluebell
Celandine	Box	Mustard
Coltsfoot	Coltsfoot	Broom
Crocus	Crab apple	Forget-me-not
*Elm	Currants	Gorse
Gorse	Dead nettle	Holly
*Hazel	Gooseberry	Horse chestnut
*Poplar	Gorse	*Oak
Prunus	Laurel	*Plantain
Snowdrop	Maple	Thrift
Violet	Willow	Wallflower
Willow		
Winter aconite		
*Yew		

June	July	August	Sept./October
Bindweed	Bergamot	Balsam	Balsam
Catmint	Borage	Borage	Gorse
Charlock	Cornflower	Chicory	Ivy
Cranesbill	Cranesbill	Golden rod	Michaelmas
Jacob's ladder	Figwort	Mallow	daisy
Lucerne	Flax	Marjoram	Sea lavender
Melilot	Hogweed	*Meadowsweet	Thistle
Firethorn	Hollyhock	Mint	
Red clover	Hyssop	Mullein	
Thyme	Knapweed	Sunflower	
Vetch	Lavender	Purple	
Viper's	Mallow	loosestrife	
bugloss	*Meadowsweet	Sage	
White byrony	Mignonette	Sea lavender	
*Wild rose	Mullein	Thistle	
	*Poppy	White	
	Privet	charlock	
	Ragwort		
	Red clover		
	Sage		
	Sweet		
	chestnut		

Appendix 3

BRITISH LEGISLATION AFFECTING BEEKEEPERS

Agriculture (Miscellaneous Provisions) Acts 1941 and 1954
Bees Act 1980
Weights and Measures Act 1963
Food and Drugs Act 1955
Honey Regulations 1976
Trade Descriptions Act 1968 and 1972
Consumer Safety Act 1978
The Glazed Ceramic Ware (Safety) Regulations 1975
The Foul Brood Diseases of Bees Orders
Importation of Bees Order 1980

Many of the Acts which affect beekeepers are drawn with wide powers which enable Ministers to make Orders under the Acts to meet specific circumstances as and when they arise. The list given above is by no means complete, but gives guidance as to the main body of legislation.

The finer points of the impact of the law on beekeeping is a matter for the legal specialist who is also a beekeeper and no attempt will be made to go into the niceties here. However, some of the more important points are worth noting.

At the moment no licence is needed for the keeping of bees and there are no specific requirements as to where hives should or should not be sited subject to the overriding consideration of 'nuisance'. In simple terms this means that the beekeeper must not keep his bees in a situation or in a manner such that a near neighbour is denied the normal enjoyment of his house and garden.

Honey Offered for Sale

There are a number of particular requirements and definitions which apply to honey offered for sale and the most important of these are listed below.

Quantities

When prepacked, quantities of more than $\frac{1}{2}$ oz net weight should be one of the following: 1 oz, 2 oz, 4 oz, 8 oz, 12 oz, 1 lb, $1\frac{1}{2}$ lb, or multiples of 1 lb. If sold loose, i.e. from a bulk container, it must be sold by net weight and the unit price displayed. Comb and chunk honey may be prepacked in any weight.

Labelling

Labels on honey containers must show the name and address of the producer, the packer or the seller. The net weight must be shown in both imperial and metric units in upright plain lettering (sans serif) 2mm high for quantities of 1 oz (28g), 3mm high for 2 oz (57g) or 4 oz (113g), 4mm high for 8 oz (227g), 12 oz (340g), 1 lb (454g), $1\frac{1}{2}$ lb (624g), or 2 lb (907g), and 6mm high for quantities over 2 lbs.

The lettering must be against a contrasting background and the units of weight should be at least half the above heights. The imperial and metric weights should be two type spaces apart and there should be nothing between them. There should be one type space between the figures and the units.

Description

The contents may be described as 'honey', 'comb honey', 'chunk honey', 'honeydew honey', 'baker's honey', or 'industrial honey'. A regional description is permitted, e.g. 'Somerset Honey' and so is a reference to a floral source such as in 'Heather

Honey', or 'Clover Honey', although in this case the honey must be predominantly from the floral source indicated.

The use of the word 'pure' raises difficulty over its legal definition and is best avoided.

Mis-description

It is an offence to use descriptive words or pictures which are untrue or misleading. The most obvious case of this would be imported honey repacked and sold as, say, Somerset or English honey.

Care is needed in choosing the design for a honey label. The inclusion of artists' sketches of flowers could be regarded as misleading. It is unsafe to incorporate any wording or pictorial representation unless the contents of the jar correspond.

Composition of Honey

No substances other than honey may be added. The honey should be free from moulds, insect debris or other organic or inorganic matter. Honeydew honey or a blend of honeydew honey should have an invert sugar content of not less than 60 per cent and a sucrose content of not more than 10 per cent. The water content should be less than 25 per cent. There are permitted maximum and minimum levels of hydroxymethyl-furfural (HMF) and diastase.

The ordinary beekeeper will not have the laboratory facilities to test his honey for these levels, but he need not concern himself unduly. Only the application of heat for a considerable time is likely to result in the honey falling outside the permitted levels, and if it is processed and packed as we have suggested in this book, no trouble is likely to arise on this score. There is a faint possibility that the HMF figure might rise if honey is kept in store in warm conditions for a long time—and by this we mean years. They are included in this brief survey to show the heights of fancy to which our masters in Brussels can rise when they really try.

Most of the regulations have arisen through British Acts of Parliament passed for the protection of the consumer and others arise from the Recommended European Regional Standard for Honey drawn up by the Codex Alimentarius Commission of the

World Health Organization Food Standards Programme. The Honey Working Party has added its quota and no doubt the regulations will proliferate as Parkinson's Law gathers momentum.

Accepted Definitions

HONEY. The fluid or crystallized food which is produced by honeybees from the nectar of blossoms or from secretions found on living parts of plants other than blossoms which honeybees collect, combined with substances of their own and left to mature in honeycomb.

COMB HONEY. Honey stored by honeybees in the cells of freshly-built broodless combs.

CHUNK HONEY. Honey which contains at least one piece of comb honey.

BLOSSOM HONEY. Honey produced wholly from the nectar of blossoms.

HONEYDEW HONEY. Honey produced wholly or mainly from the secretions of or found on living parts of plants other than blossoms.

DRAINED HONEY. Honey obtained by draining uncapped broodless honeycomb.

EXTRACTED HONEY. Honey obtained by centrifuging broodless honeycomb.

PRESSED HONEY. Honey obtained by pressing broodless honeycomb.

Further Reading

Anatomy
The Anatomy of the Honeybee, R. E. Snodgrass (Cornell U.P., 1975)
Anatomy and Dissection of the Honeybee, H. A. Dade (Inter. Bee Research Assoc., 1978)
Honeybee Anatomy, British Beekeepers Assoc.
Insect Flight, J. W. S. Pringle (Oxford U.P., 1975)

Behaviour
Bees: Their Vision, Chemical Senses and Language, K. von Frisch (Cornell U.P., 1972)
Insect Natural History, A. D. Imms (Collins, 1971)
The World of the Honeybee, C. G. Butler (Collins, 1975)

Diseases
Diseases of Bees, Ministry of Agriculture Bulletin No. 100 (H.M.S.O.)

Flowers and Bee Forage

Illustrated Guide to Pollen Analysis, P. D. Moore and J. A. Webb (Hodder, 1978)

Pollen Loads of the Honeybee, D. Hodges (Inter. Bee Research Assoc., 1974)

Trees and Shrubs Valuable to Bees, M. F. Mountain (Inter. Bee Research Assoc., 1975)

Plants and Beekeeping, F. N. Howls (Faber and Faber, 1979)

Practical Beekeeping and Equipment

Beekeeping, Ministry of Agriculture, Bulletin No. 9 (H.M.S.O.)

Complete Handbook of Beekeeping, H. Mace (Ward Lock, 1976)

Manual of Beekeeping, E. B. Wedmore (Bee Books, 1975)

Queen Rearing

Queen Rearing, H. H. Laidlow and J. E. Eckert (Univ. California Press, 1962)

Swarming

Swarming of Bees, Ministry of Agriculture, Bulletin No. 206 (H.M.S.O.)

Journals

Beecraft, 15 Westway, Copthorne Bank, Crawley, Sussex (monthly)

British Bee Journal, 46 Queen Street, Geddington, Kettering, Northants (monthly)

Most aspects of beekeeping are covered by Ministry of Agriculture advisory leaflets available from H.M.S.O.

Useful Addresses

(It is courteous to send a stamped addressed envelope when asking for information—it speeds a reply too!)

The British Beekeepers Association: The Secretary, National Agricultural Centre, Stoneleigh, Kenilworth, Warwickshire.

The International Bee Research Association: The Secretary, Hill House, Chalfont St Peter, Buckinghamshire, SL9 0NR

The National Beekeeping Specialist, Luddington Experimental Horticulture Station, Stratford-on-Avon, Warwickshire, CV37 9SJ

Appliance Manufacturers

Robert Lee Ltd., Beehive Works, George Street, Uxbridge, Middlesex

Steele and Brodie Ltd., Wormit, Fife, Scotland

E. H. Taylor Ltd., Beehive Works, Welwyn, Hertfordshire, AL6 0AZ

E. H. Thorne (Beehives) Ltd., Beehive Works, Wragby, Lincolnshire, LN3 5LA.

Ministry of Agriculture, Fisheries and Food (Publications), Tolcarne Drive, Pinner, Middlesex, HA5 2DT

National Honey Show: The Secretary, Gander Barn, Southfields Road, Woldingham, Surrey.

The National Beekeeping Centre, Stoneleigh, Kenilworth, Warwickshire.

Samples of bees for examination for suspected disease should be packed in a matchbox (about 30 bees) and sent to The National Beekeeping Specialist at Luddington.

Glossary

Aberration, chromatic	Formation by a lens of coloured fringes round the image. The effect is due to glass having different refractions for light of different colours.
Acarapis woodi	A mite which infests the main tracheae of bees. The causative agent of acarine disease.
Aorta	The large blood vessel leading from the heart to just behind the brain, where it discharges the blood supply.
Apodemes	Thickened ridges on the body plates, either for strength or for the attachment of muscles.
Aplanat lenses	A lens consisting of two or more components, fixed in a mount, which will produce a flat image. They are usually computed to correct chromatic aberration too.
Arolium	The part of the foot which enables the bee to cling to smooth surfaces.
Auricle	A sloping shelf on the basitarsus of the hind leg where pollen is collected before being pressed up into the corbicula.

Bacillus larvae	A spore-forming bacillus, the causative agent of American Foul Brood.
Basitarsus	The fifth joint of the leg (counting from the body).
Braula coeca	A species of diptera. An external parasite of bees, especially queens.
Brood	A term used to cover all the juvenile forms from newly-hatched larvae to sealed pupae.
Cardines	The hinge-like chitinous processes by which the proboscis is slung below the head.
Chitin	The hard material which comprises the horny skeleton and which also appears in other anatomical parts as strengthening struts etc. It is extremely resistant to decay and decomposition and in this respect is only equalled by pollenin, the hard outer casing of pollen grains.
Chloro-benzilate	An acaricide used in Folbex strips for the treatment of acarine disease.
Colony	A 'family' of bees, i.e. workers, drones and a fertile queen. Also called a stock. A hive is the man-made construction in which a colony lives.
Corbicula	Literally 'little basket'. The apparatus on the tibia of the hind legs which enables loads of pollen to be carried.
Cuticle	The outer, chitinous, skeleton.
Cytochrome	A cell constituent which enables atmospheric oxygen to be used in muscular effort. Wing muscles are particularly rich in it.
Cytology	The study of biological cells.
Dehiscence	The liberation of ripe pollen from the anthers of flowers. This will often occur at different times of the day for different species.
Dextrose	A monosaccharide. Same as glucose and grape sugar.
Disaccharides	Sugars commonly occurring in nature, e.g. sucrose, the common table sugar. The difference between disaccharides and monosaccharides is one of molecular arrangement; for instance, the formula for the disaccharide

sucrose is $C_{12}H_{22}O_{11}$ and that of the mono-saccharide dextrose is $C_6H_{12}O_6$. Polysaccha-rides are even more complex but still consist of arrangements of carbon, hydrogen and oxygen atoms.

Doublet lenses   Partially corrected lenses to give a flat field. See also under *Aplanat lenses*.

Diatomaceous earth   A filtering agent. Diatoms are the silica skeletons of microscopic unicellular organisms.

Endophallus   The genital apparatus of the drone. It is normally carried within the body but is everted at the moment of copulation.

Fixation   A process of arresting decay and leaving bio-logical material in a condition as nearly as possible to what it was in life.

Enzymes   Catalysts, i.e. substances which, in minute amounts, instigate and promote chemical changes without themselves being used up in the process.

Exoskeleton   The chitinous outer covering.

Flagellum   The distal portion of the antenna. It is jointed and very rich in sense organs.

Formol alcohol   A preserving fluid for specimens.

Fructose   A monosaccharide. Same as levulose. See also under *Disaccharides*.

Glucose   Same as dextrose.

Glycerine jelly   A mountant for delicate dissections, pollen etc.

Glycogen   A polysaccharide built up of numerous dextrose molecules. It is stored in the body by bees as a source of future energy.

Haemocyanin   A blue/green substance in the blood of bees. A vehicle for the transport of oxygen throughout the body. Analagous to haemoglobin in mammals.

Hamuli   Small hooks on the leading edges of the hind wings which, in flight, engage with a fold in the trailing edge of the front wing.

Histology   The study of biological tissues.

Honey sac   A small sac at the posterior end of the oesophagus and just before the ventriculus. Nectar is stored here by the forager on the flight home. The enzyme diastase is added here to initiate conversion into honey. Sometimes called the 'honey crop'.

Hormones   Substances secreted by glands. They circulate within the body of the organism producing them and effect profound changes in functions.

Hypopharyn-   Glands in the heads of bees of nursing age which
geal glands   secrete brood food.

Hygro-   The attraction of a substance for water, e.g.
scopicity   honey is hygroscopic and will absorb water if left exposed.

Intersegmental   Thin elastic membranes between the hard
membranes   segments of the body. These allow movement and prevent the body being a rigid box.

Invertase   An enzyme secreted by workers. Its purpose is to break down complex into simple sugars.

Levulose   A monosaccharide. Same as fructose and fruit sugar.

Macerating   The dissolving out of soft tissues in a dissection in order better to see constructional details of the harder tissues.

Malpighian   The 'kidneys' of an insect. They are fine tubes
tubules   which lie in the body cavity and extract waste products from the blood and pass them down to the excretory system.

Malpigh-   A spore-forming protozoan. The causative
amoeba   agent of Amoeba Disease.
mellifica

Mandibles   The 'jaws'.

Mass   Providing larvae with a mass of food more than
provisioning   is sufficient for its immediate needs. Worker larvae are fed in this way for the first few days and queen larvae for the whole of their larval life.

Melissopalyn-   The microscopical examination of the solid
ology   particles in honey.

Micropyle	The pore in the end of an egg through which sperms enter for fertilization. In pollen grains, the openings through which pollen grains will grow down the stigmas of flowers.
Mono-saccharides	Simple sugars, e.g. levulose and dextrose. See also under *Disaccharides*.
Nectar	The sugary solution in water processed by plants from plant sap.
Ocelli	The three simple eyes on the top of the bees head.
Octoacetyl sucrose	A 'denaturing' substance added to sugar intended for agricultural use etc. in order to make it unsuitable for human consumption. It is extremely bitter, but acceptable to bees.
Ostia	Valves along the heart enabling blood to be drawn from the body cavity and directed forward by peristaltic constrictions of the heart with co-ordinated opening and closing of the ostia.
Peristalsis	A wave-like motion of constriction passing along a tube, e.g. along the gullet in the act of swallowing.
Pheromones	Secretions of glands used externally. They elicit reactions in other individuals of the same species.
Phragmae	Thickened ridges on the internal edges and sides of body segments for strengthening and attachment of muscles.
Plumose	Branched like a feather. A description of the hairs on a bees body.
Proboscis	The 'tongue' parts comprising a number of separate pieces. Used for sucking up liquids.
Proctiger	The vestigial last abdominal segment. It carries the anus.
Prothorax	The anterior part of the thorax.
Ramus	The curved extension of the sting lancet. It is articulated to the triangular plate and is part of the driving mechanism of the sting.
Rastellum	The 'rake' of teeth attached to the inside lower

	edge of the tibia. It is used to scrape pollen off the hairs of the basitarsus.
Rhabdon	Long transparent rods reaching from the base of the crystalline cone behind each facet of the compound eye back to the retina. Each rhabdon is surrounded by an opaque pigment.
Scape	The rigid rod-like part of the antenna nearest the head.
Scutal fissure	A flexible 'gusset' in the thorax which enables the box-like thorax to distort in flight.
Seminal vesicles	Sacs in the drone in which are stored matured spermatazoa.
Spermatheca	A spherical receptacle in the queen in which sperm is stored after successful mating.
Spiracles	Breathing orifices down the sides of insects.
Sternites	The ventral plates of the exoskeleton.
Stock	The same as 'colony'.
Stylets	The upper parts of the sting lancets.
Sucrose	A disaccharide. Table sugar.
Streptococcus pluton	A spore-forming micro-organism. The causative pathogen of European Foul Brood.
Supers	Boxes placed on a colony, usually above a queen excluder, for the collection and storage of honey by the bees.
Tergites	The dorsal plates of the exoskeleton.
Tracheae, tracheoles	Tubes leading from the spiracles carrying inhaled air. Tracheoles are small tracheae.
Varroa jacobsonii	A parasitic mite of the order Arachnida. An Asiatic migrant. The causative agent of varraosis.
Ventriculus	The stomach. The site of all the processes of digestion.
Vasa deferentia	The tubes connecting the testes of the drones to the seminal vesicles.

Index

Abdomen, of bee, 105
Acarine;
  cause, 204
  diagnosis, 206
  effects, 205
  Frow treatment, 207
Addled brood;
  cause, 200
  cure, 200
Agricultural Development
  and Advisory Service, 213
American foul brood;
  cause, 196
  destroying colony, 197
  symptoms, 197
Amoeba disease;
  cause, 204
  treatment, 204
Antennae, 99, 103
Apiaries, 10, 15
  sites, 34

Arrival of nucleus,
  procedure, 21-3

Bailey, Dr L., 210
Basic Beekeeping, 7
Bee 'dances', 88
Beekeeping associations, 9, 17
Bee senses;
  hearing, 69
  taste and smell, 68, 87
  vision, 67, 87, 98, 99
Bee space, 53
Beeswax;
  formation, 135
  uses, 136
Blood and circulation, of bee,
  110
Brain, of bee, 99, 100
Braula coeca;
  definition, 204
  removal, 204

Breeding bees, 19, 20
Brood box, 22
    chamber, 51, 52, 53
Brood food theory
    (Gerstung's theory), 146
Brood-rearing, 77, 78
Buckfast strain, 19
    system, 36, 37
Bulman, Michael, 126

Carr, William Broughton,
    see W.B.C. hive
Carson, Rachel, 213
Cell;
    building, 75
    capping, 76
    shape, 75, 76, 79
Chalk brood;
    cause, 199
    precautions, 200
Chemical repellants;
    applications, 173
    benzaldehyde, 174
    carbolic, 174
    proprionic anhydride, 174
Chilled brood, 201
Clearer boards;
    Canadian board, 175, 178,
        179
    porter escape, 175, 178
Protective clothing, 10-13
Clustering, 71, 111
Colony;
    growth, 29-31
    handling when 'touchy',
        65-6
*Complete Instructions for
    Beekeeping*, 139
Compound eye, of bee, 98,
    100

Cornea, mounting, 227, 228
Courses, for beekeeping, 9
Cover cloths, 64

Dade, H. A. 8, 218
Digestive system, of bee,
    107-110
Dissection;
    of thorax, 224-5
    preparation of
        exoskeleton, 228-9
    technique, 223-4
    tools, 221-2
Distal end, definition, 103
Division of labour, in
    colony, 83-5
Drone, 19, 83, 113-15
    comb, 77
Dry cells, 226-7
Dummy frames, 22-3

Eggs, 113
Electro-magnetic effects, on
    bees, 70
European foul brood;
    cause, 198
    symptoms, 198
    treatment, 199
Exoskeleton, of bee, 98
Extraction of honey;
    hygiene, 181
    Pratley uncapping tray,
        183
    radial type extractor, 186-7
    tangential type extractor,
        185
    uncapping, 182, 183

Fanning, 73
Feeding process, 72-3

Fixation, 226
'Flower honey', 118
Folbex strips, 208
Foragers, 30, 31, 33, 37, 72
Forced air blowers, 174
'Forest honey', 124
Frames;
  inspection of, 24, 64, 65
  propolized, 63, 65
Frame spacing;
  castellated runners, 57
  metal ends, 56
  panel pins, 57
  self-spacing Hoffman, 55
  self-spacing with screw
    eyes, 55
  Yorkshire spacers, 56
Frisch, Karl von, 88

Gloves, 11
Glycerine jelly, 230
Gorging, 71

Handling bees;
  annoying, 61, 62, 64
  controlling, 61
  favourable conditions, 62
  planned procedure, 63, 64
  techniques, 59-61
  'touchy' colony, 65, 66
Head, of bee, 98
Hives;
  choosing, 15-17, 45, 48
  consistency, 34
  double-walled, 46-8
  making your own, 46
  preservation, 53, 54
  primitive, 40, 41
  single-walled, 46-8, 50
  tools, 13, 63

Hodges, Dorothy, 133
Honey;
  contents, 118
  crystallization, 119-20
  filtering, 120, 121
  medicinal properties, 125,
    126
  storing conditions, 119
Honeydew, 123-4
  honey, 124
Honey, examination of;
  centrifuge tubes, 129-31
  pollen identification, 132,
    133, 135
  preparing a slide, 128-9
  reasons for study, 127
  sedimentation, 132
Howes, F. N., 123

International colour code, 26
Isle of Wight disease, 209

Jansha, Anton, 139

Killing bees, 225

Langstroth, L. L., 45
Larvae, 72, 74
Legs of bees;
  front, 102
  hind, 103-4
  middle, 103

Mating, 115, 116
  'sign', 116
Microscopy;
  illumination, 220, 221
  prismatic microscope,
    219-20
Ministry of Agriculture, 195

Nasonov gland, 105
National Agricultural
  Advisory Service, 94
Nectar, 108, 117
  flow, 33, 72, 78, 127
Nectar sources;
  Bell Heather, 122
  Blackberry, 123
  Clover, 121
  Field beans, 123
  Hawthorn, 123
  Heather, 122
  Lime, 122
  Sainfoin, 123
  Sycamore, 123
Neglected drone brood;
  cure, 201
  stunted drones, 200
Nest requirements, 39
Nosema disease;
  cause, 201
  symptoms, 202
  treatment, 202, 203
Nucleus, 11, 12
  box, 21
Paralysis;
  cause, 210
Pheromones, 90
Plants and Beekeeping, 123
Pollen, 107, 108, 109, 140
*Pollen Loads of the
  Honeybee*, 133
Pollination, 93-5
Propolis, 11, 13, 76, 138, 139
'Proximal' end, definition of,
  103

Queen;
  characteristics, 23, 24, 27

clipping, 27
finding, 23, 24
handling, 26
marking, 26, 27
recognition, 81, 82
Queen-excluder, 24
Queenlessness, 82
Queen-rearing;
  for small-scale beekeeping,
    162
  'grafting' method, 163-6
  mating virgin queens,
    166-8
Queen-right, 65, 71
Queen substance, 82, 84, 90,
  145

Reproductive organs, of bee,
  111-14
Respiratory system, of bee,
  110, 111
Royal jelly, 140-1

Sac brood
  cause, 199
  re-queening, 199
Scout bees, 87
Silent Spring, 213
Smokers, 22
  bent-nosed, 15
  straight-nosed, 15
Smoking the colony, 66
Spray and pesticides, 212,
  213, 214
Spring honey;
  rape combs, 172
  Shake and brush method,
    171, 172
Sting, 11, 60, 105, 107

Storage of honey;
   direct bottling, 193
   method, 190-1
   warming box, 191-3
Straining honey;
   equipment, 188-9
   method, 189-90
Strains available;
   Carniolan, 18
   Caucasian, 18, 78
   Italian, 18, 78
   North European, 18
Source of bees, 20-1
Supersedure, 26, 27, 149
Swarm, artificial, 155
   examining for, 31, 35,
     155-8
   origins, 144
Swarm control, 31, 149, 150
   Demaree method, 150, 152
   Pagden method, 150
   Snelgrove method, 153,
     154

Swarming, reasons for, 83,
   146-8

Temperature, of nest, 74, 75
'Tree honey', 118

Varroasis;
   cause, 210
   importing bees, 212
Venom, 107, 141, 142
   gland, 107

Water supply, 28, 34
'Wax mirrors', 135
Wax moth, 215
Wax scales, 75, 76
W.B.C. (William Broughton
   Carr) hive, 49-50, 53
Wings, of bee, 100-2
Workers, 79, 80

DATE DUE

GAYLORD			PRINTED IN U.S.A.